Numerical Control

WILEY SERIES IN COMPUTING

Consulting Editor

C. A. Lang, *Computer Laboratory, Cambridge University*

Numerical Control—Mathematics and Applications:
P. Bézier,
Professeur au Conservatoire National des Arts et Métiers
and
Technological Development Manager, Renault, France

Some shapes created using numerical control

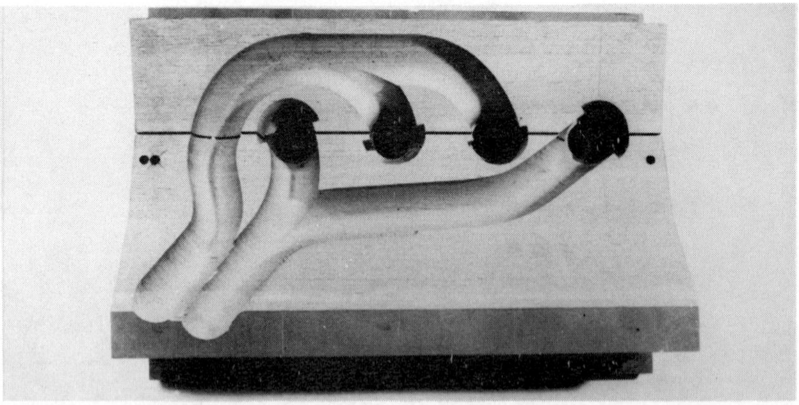

An exhaust-manifold pattern defined by UNISURF curves and cut by a spherical cutter

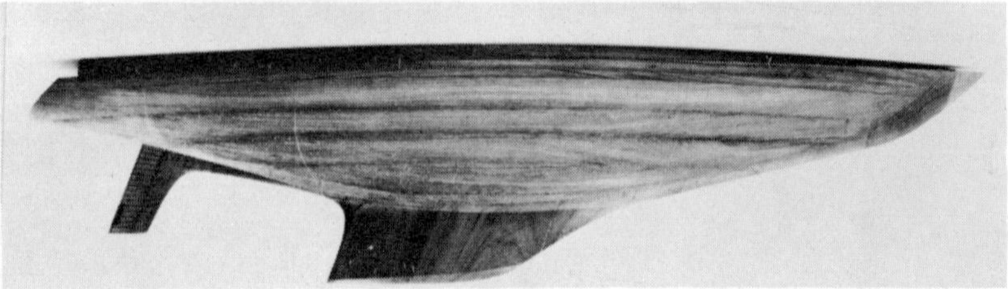

A model yacht hull made using the UNISURF system

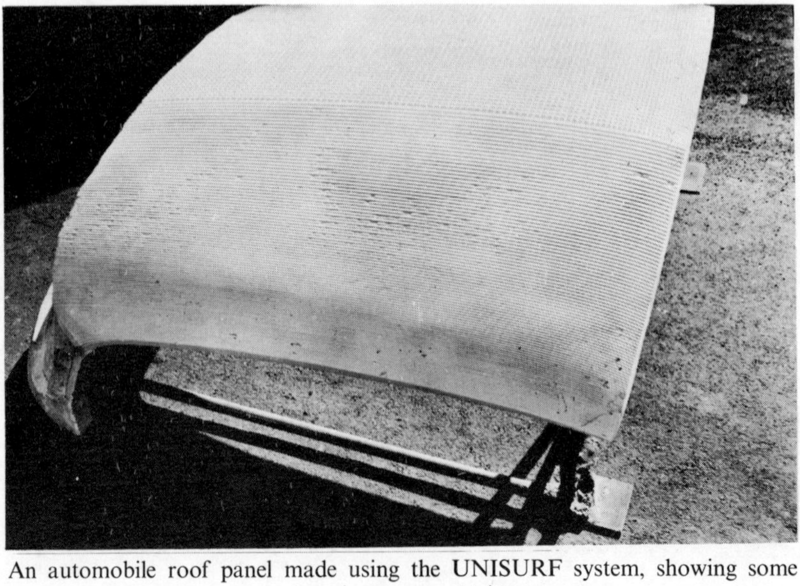

An automobile roof panel made using the UNISURF system, showing some conjugate generatrices

Numerical Control

Mathematics and Applications

P. Bézier

Professeur au Conservatoire National des Arts et Métiers
and
Technological Development Manager, Renault, France

Appendixes by

A. R. Forrest

Computer Laboratory, Cambridge

Translated by

A. R. Forrest and Anne F. Pankhurst

JOHN WILEY & SONS

London · New York · Sydney · Toronto

First published under P. Bézier. *Emploi des Machines à Commande Numerique*
by Masson & Cie, Editeurs, Paris 1970.

Library of Congress catalog card No. 70-39230

ISBN 0 471 07195 1

Set on Monophoto Filmsetter and Printed by J. W. Arrowsmith Ltd., Bristol, England

Preface

Professor Bézier's book, like others already available in English, deals with the design of numerically controlled machine tools and their uses. Unlike the other books, however, it combines this with a review of mathematical techniques for describing shapes to be produced by numerically controlled machines. This has provided the main incentive to publish an English edition.

Up to the present time, the most common use of mathematical techniques in numerical control has been within the various numerical-control language systems, whereby the shape of a component designed in some undefined way may be coded, and the necessary tool path computed. Such techniques must be suitable for the two roles of shape description and tool-path computation. A more recent development now starting to appear on the industrial scene is that of computer-based systems for designing mechanical components. The end result of these systems is a numerical description of a component based on a mathematical formulation that is stored in a computer.

In a design system, the mathematical methods must have additional properties. They must be able to provide a means whereby the designer can create and modify a shape in a natural and easy manner, and they must be suitable for any analytical computations (such as weight, centre of gravity, or stress) that may be required as part of the design system. With a computer-stored description of the shape, production on a numerically controlled machine tool is the natural method. Ideally, the cutting information would be generated automatically from this shape description and a knowledge of the machine to be used, but, in practice, a human must add more information pertinent to the production operation. In his intensely practical UNISURF system and the mathematics on which it is based, Professor Bézier has made a particularly valuable contribution with a system for the design and cutting of three-dimensional curved shapes. This is but one of the mathematical techniques described in this book.

The use of numerically controlled machine tools is now firmly established. Design and production systems based on an increased armoury of mathematical techniques for manipulating shapes by computer are likely to increase greatly the use and effectiveness of numerically controlled machine tools from now on.

C. A. LANG
Leader, Computer-Aided Design Group,
University of Cambridge

Contents

Introduction

In the course on mechanical manufacture given at the Conservatoire National des Arts et Métiers, the use of numerically controlled machines has an important place.

Numerical control is of recent origin. Its evolution has been rapid because it makes use of, in particular, the latest advances in electronics and computer technology. There are innumerable published articles on this subject and it is difficult for the layman to build up a complete picture from them.

The construction of these machines and their control systems, and, similarly, the elaboration of their languages, demands knowledge in many fields: mathematics, electricity, information theory, automation, pneumatics, hydraulics, not to mention mechanics. This book is aimed at those whose work leads them to make use of such equipment. It is not necessary for them to know all these sciences in depth, but, certainly, sound basic ideas are indispensible to those who wish to make the best use of complex and highly automated plant. The reader will find it of great advantage to consult textbooks, such as the treatises of M. l'Ingénieur General Salmon,[1] Professor W. Simon,[2] M. Jacques Thilliez[3] and others,[4,5] in which the control systems, the machines and the languages are described.

In this work, many examples, methods and names of machines have been quoted. As they belong to a particularly vigorous industry, one must not regard the information as unchanging and definitive.

All those readers whose work is directly tied to the utilization of numerical-control processes, and who wish to keep informed of their evolution, must study the numerous articles published in the technical and foreign Press.

Need I add that, in the preparation of this book, I have been helped by many colleagues who have been willing to inform me of the results of their work or who have directly helped me in the accomplishment of my task?

They are too numerous to be mentioned individually, but I should especially like to express may deepest thanks to MM. Lagrange, Fririon, Riaux, Casselli, de Vanssay, Chillon and Hamon, who have given me invaluable help. I am indebted to Dr. Forrest and Mr. Lang for their ready exchange of information, as well as for their work with Mrs. Pankhurst in the translation of this book.

To them all I express my sincere gratitude.

P. BÉZIER

References

1. P. Salmon, *Introduction to the numerical control of machines* (*Initiation à la commande numérique des machines*), Société des Publications Mécaniques.
2. W. Simon, *The numerical control of machine tools* (*Commande numérique des machines-outils*), Éditions Eyrolles, 1967.
3. J. Thilliez, *The numerical control of machines* (*La commande numérique des machines*), Éditions Dunod.
4. G. G. Ertell, *Numerical control*, Wiley–Interscience, 1969.
5. N. O. Olesten, *Numerical control*, Wiley–Interscience, 1970.

Chapter 1

Generalities

1.1 HISTORY

The first numerically controlled machines appeared during the 18th century: the Falcon (1728) and Jacquard (1800) looms. For 150 years their only descendants were piano players or barrel organs.

About 1942, the Bendix Corporation had to solve a problem posed by the manufacture of the three-dimensional regulator cam of an injection pump. The conditions to be met were so numerous that it was decided to entrust the computations to an automatic machine, and to use the results to control directly the movements of a special machine.

Almost simultaneously, the Massachusetts Institute of Technology, in collaboration with I.B.M. and the Cincinnati Milling Company, was completing a similar exercise.

The use of this method was, naturally, somewhat limited, because the method was applicable to precision components with very complicated shapes, whose cost was, in any case, very high.

Nearly ten years passed before the appearance, in the U.S.A., of machines that were capable not only of milling complex surfaces, but of drilling, tapping and boring.

Manufacturers for some time adopted a somewhat reserved attitude to this technique, and there were few orders; but the American Government had not overlooked the economic and strategic benefits of this technique and ordered 200 machines which they placed at the disposal of industries in the public sector and certain of their contractors.

The impulse having been given, the number of numerically controlled machines multiplied rapidly, and the trend spread to other countries. Some relevant statistical details are given in Chapter 7.

Since 1960, we have seen the development, on a new basis, of milling machines which are proving valuable in the aerospace industry in particular, and in other areas such as the manufacture of tools for metal stamping and forging.

Thus the evolution of numerically controlled machines may be divided into three distinct periods:

1. From 1942 to 1945: the appearance of the first small- or medium-sized contour-milling machines.
2. From 1950 to 1960: the appearance and development of 'point-to-point' drilling/boring machines of all sizes.
3. After 1960: the simultaneous development of 'point-to-point' and contouring machines, the former being, nevertheless, in the large majority.

The technologists solved the very difficult problem of contouring before attacking the much simpler problem of point-to-point working, while the natural progression goes from the simple to the complex.

The only plausible explanation for this phenomenon is that the former problem arose during wartime, and cost was not very important compared with precision and lead time.

To make point-to-point working profitable, machines specially adapted to this mode of working had to be developed; it has also been necessary for the control-systems technology to overcome the initial difficulties, and for the price and reliability to make the improvements generally experienced in the early years of any new technology.

1.2 FUNDAMENTAL PRINCIPLES

In the generally accepted convention, a machine tool is said to be 'numerically controlled' if it works in an automatic or semiautomatic manner according to instructions transmitted to it in a coded form.

This definition immediately calls for two comments:

(a) The symbols by means of which the instructions are expressed can be numerals, letters and various punctuation marks (brackets, commas, asterisks, slashes, etc.). Strictly, the term 'symbolic' is more appropriate than 'numerical', but we have retained the latter term which was, originally, justifiably used to describe simple systems for which the ten decimal digits sufficed to define all the essential data.

(b) Instructions were first given in numerical form when dimensions were placed on a drawing, or a speed was defined by a number of revolutions per minute, by the selection of a gear shift or by the position of a gear lever.

The fundamental characteristic of numerical control is the automation or semiautomation of the execution of instructions, and not solely the form, numerical or symbolic, in which they are expressed.

Using the vocabulary of automation, an input is represented by an instruction, the output is that given by the corresponding transducer, and their difference is the error signal, i.e. the main data of the servo loop.

The fundamental input values concern Cartesian or polar coordinates, to which, sometimes, are added the velocities of translation or rotation. During the running of the machine, the feedback loops may take into account the acceleration values supplied by the appropriate transducers.

Lastly, one can foresee a still more complete optimization of the machines, taking into account the frequencies and magnitudes of vibration of certain mechanisms, the constraints or the strains, the temperature of the cutting edge, etc. This concept is known as 'adaptive control'.

Bibliography

1. M. M. Chalvet, 'Human aspects and economical consequences of n.c. (Aspects humains et conséquences économiques de la C.N.)', Société des Ingénieurs de l'Automobile (S.I.A.), lecture series on numerical control (May 1963–March 1964).
2. M. R. Prudhomme, 'Industrial applications of n.c. (Applications industrielles de la C.N.)', Société des Ingénieurs de l'Automobile (S.I.A.), lecture series on numerical control (May 1963–March 1964).
3. M. M. Chalvet, 'The development of n.c. for machine tools (Développement de la C.N. des machines-outils)', *Ingénieurs et Techniciens*, March 1966.
4. F. Mitthof, 'Numerik 1965. Survey of the world's numerically controlled machine tools. Part 1 : Europe (Uebersicht der numerisch gesteuerten Werkzeugmaschinen der Welt. Teil 1 : Europa)', *T.Z. für praktische Metallbearbeitung*, No. 8, August 1965.
5. 'N.C. What it means to metalworking', *American Machinist*, 25 October 1954 (McGraw-Hill).
6. 'N.C. The second decade', *American Machinist*, 26 October 1964 (McGraw-Hill).
7. Special number of *Production* (Bramson), December 1964.
8. Special number of *Production* (Bramson), December 1966.
9. 'Whence the N.C. evolution', *The Iron Age*, 1 June 1967.
10. J. E. Standford, 'N.C. Good today, great tomorrow', *The Iron Age*, 7 September 1967.

Chapter 2

Control Systems

There are a very large number of control systems; the reader will find only general outlines here. He should refer, for fuller details, to works on automation, mechanics, hydraulics, electricity, machine construction, etc.

The principal components of a control system are:

(a) The means for transmission of data (or the input values).
(b) The servo systems.
(c) The motors.

All must satisfy the conditions imposed by the execution of the work cycles.

2.1 WORK CYCLES

Point-to-point working

Consider a numerically controlled machine, in the form, for example, of a column drilling machine with a cross-slide table (Figure 2.1).

On the completion of a drilling operation, the drilling head is returned to its raised position, and the tool is then some distance above the workpiece.

Before performing another work cycle, the table must be moved to another position defined by a pair of Cartesian coordinates. This displacement is produced by the movement of the table on the saddle, and of the saddle on the frame; with some systems these motions are performed in sequence, simplifying the apparatus, but slowing the work rate. Thus, in nearly all cases, the two movements are started simultaneously, each stopping when the corresponding error signal becomes less than the preassigned limit.

The two movements, therefore, are independent, and the form of the trajectory from one work position to another is of no importance, provided that the terminal point is exact.

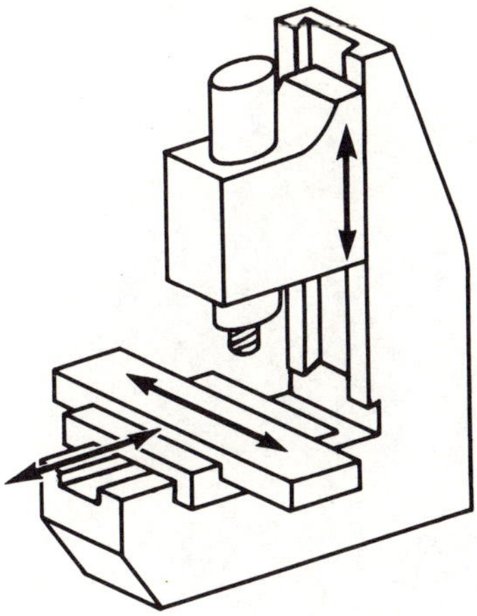

Fig. 2.1

When both these displacements are completed, the tool performs its work, combining a rotation about its axis with a series of translations parallel to its axis, in sequence, corresponding respectively to the stages of fast traverse, work feed and rapid return.

Straight-cut milling

Numerical control can be used to mill, on a workpiece, a number of faces perpendicular to one of the three orthogonal axes of the machine.

This mode of working differs a little from point-to-point working:

1. The displacements perpendicular to the axis of the tool head are not simultaneous, at least not during milling.
2. These displacements take place with a speed equal to the work feed and not at the rapid traverse speed.
3. Their amplitudes take into account simultaneously the dimensions of the workpiece and of the tool.
4. No movement may, at any time, overshoot the assigned end point, where, in point-to-point working, there may be a 'balancing' around the final position.

There are some machines, which appeared before numerically controlled machines, which perform the same kind of work, but whose mode of operation is based on a different principle. The end of each movement is defined by a cam acting on one of the electrical switches positioned along the slides of the machine, and a commutator selects the slide that corresponds to the phase of the cycle being executed (Figure 2.2).

Fig. 2.2. Ernault-Somua milling machine—control by 'peg board'

Numerical control has not eliminated this system, but has the advantage over it of not requiring time for setting up; for small runs this feature more than adequately compensates for the difference in capital cost, which is in favour of the cam system.

Contouring

When a large number of successive points are assigned to a machine, the movement of the tool relative to the workpiece is composed of short straight-line segments making up a path which is practically identical to a continuous curve.

The fundamental characteristic of contouring resides in the accuracy with which, at any instant, the ratios of the velocities of the different movements must be guaranteed.

On the simplest machines, there are only two simultaneous movements, but certain improved machines can control six movements at a time, some being translations and others rotations.

2.2 TRANSMISSION OF INSTRUCTIONS

Nature of instructions

The instructions necessary for the execution of each operation are not restricted to the coordinates of the points defining the movements.

Much other data can be specified, according to the degree of automation—the spindle speed, the feed rate, identification of the tool to be used, selection of the coolant, the work cycle (drilling, tapping, spot facing, etc.), the type of interpolation, etc.

Semiautomatic operation

When a task does not involve a great number of operations, and when there are only a very small number of parts to be made, automatic programming is not always economical. In such cases it is better to set down the data, either by hand or by using some printing device, on a sheet sent to the operator who feeds them, line by line, to the machine after consulting the instruction sheet which has been sent to him with the drawings of the piece to be made and complementary information concerning the resting and clamping points, the type of tools, etc.

This rule is not without exception. When a task involves a large number of similar operations, each of which takes a known time, there may be an advantage in total automation. The first advantage is to free the operator, and let him turn his attention to a complementary activity—inspection, chamfering, deburring, examination of the work data for the following job, unclamping the previous workpiece, or even cleaning—without stopping his machine, but the main advantage is to avoid the loss of attention engendered sooner or later by monotony. On a paper-mill roll or a condenser or a boiler plate, the same operation may be repeated several thousand times, and there is the risk that any error will be difficult, or even impossible, to rectify.

When semiautomatic operation has been chosen, the operator has at his disposal a control console with switches and decimal dials by means of which he enters the data.

Some systems have only binary-coded decimal scales; the operator rapidly becomes accustomed to translating a decimal number into binary-coded decimal. This results in a simplification in the control system, but this intervention of the operator in an arithmetic process gives rise to a certain mistrust among the uninitiated.

Similarly, rotary switches of the type used in telephone exchanges were used for some time, but this system appears to have lost favour.

Finally, keyboard systems similar to some desk calculating machines are sometimes used. This saves fatigue and time, but the systems are a little more expensive than other methods of input.

Automatic operation

As soon as the number of operations becomes significant, or if the run includes several parts, the automatic input of data becomes, without notable exception, the rule.

The input medium belongs to one of three different types:

(a) Punched cards.
(b) Punched tapes.
(c) Magnetic tapes.

PUNCHED CARDS

This system, once common enough, seems to have fallen out of use, except for the control of machines which assemble electronic components on printed circuits.

The execution of a program is controlled by a deck of cards, each of which corresponds to a particular operation. The advantage of this system lies in the relative ease with which an operation may be added to, or removed from, the work sequence. Nevertheless, care must be taken to arrange the cards exactly according to the numerical sequence of the operations, which is checked by a special automatic system. Therefore, after any modification a new deck of cards must be punched and numbered according to the new sequence.

PUNCHED TAPES

This is the most widely used system. The input medium is a strip 1 in wide with eight tracks which may be perforated, and a ninth track which is used by the reader drive. The spacing of the perforations is 0·1 in, and each transverse line corresponds to a character. The correspondence between the perforations and the symbols they represent is standardized. The I.S.O. code has replaced the EIA.RS.244 code.[2]

For infrequent use, paper tapes are sufficiently strong. Other tapes are made of plastics, or even of a plastics/aluminium sandwich.

Some devices scan simultaneously all the lines of a block of information corresponding to a basic operation, but the most widely used devices read the tape line by line and arrange the information in an auxiliary memory.

The reading of the tape can be accomplished by mechanical, pneumatic or optical means. The speed of reading varies between ten and 1000 characters per second.

MAGNETIC TAPES

As was implied in the previous paragraph, a punched tape contains 400 characters per metre of tape. Since contouring operations require, in general, a large amount of information, and since the speed of punched-tape readers can limit the performance of the system, some manufacturers have employed magnetic tapes whose information density is nearly one hundred times greater than that of punched tapes, and whose reading speed is also much greater.

It must, however, be noted that the use of magnetic tapes imposes several constraints. First, it is almost impossible to detect by a simple visual examination whether a tape has undergone a superficial modification which has damaged the information that it carries, and there is little chance of detecting such damage until it produces an error in the working of the machine.

Secondly, magnetic coatings attract ferrous dust; this can alter the magnetic signals, and it can damage the magnetic coating during the passage through the reading head.

For this reason, the control consoles for magnetic-tape systems are generally housed in a protective glazed area near to the machines, although the workshops in which they are to be found principally manufacture components in light alloy whose dust is practically non-magnetic.

Finally, the preparation of the tapes is executed by means of a special computer, which compels the majority of users to call on the services of an outside computer bureau.

Magnetic tapes are used less often than punched tapes, but, since about 1965, great efforts have been made to extend their use. In particular, sealed tape magazines have been developed, which protect the tapes from dust.[3–5]

Mixed operations

Certain data, such as the length of a tool, the diameter of a milling cutter or the exact position of a turning tool with respect to its carriage, are not readily known when the work program is written.

In this case, the program is transcribed onto punched or magnetic tape, and the data which are unknown until the work is carried out are given to the machine by the operator by means of switches on the control console.

Not all automatic systems are capable of taking into account the data selected at the moment of execution, and, with certain machines, the length of the tools or their diameter must be preset exactly. In other machines, the correction of the radius cannot exceed one or two millimetres.

2.3 SENSORS

Relationship with components

Servo systems must receive information regarding the instantaneous position of each moving part, to which are sometimes added the components of its velocity and its acceleration.

Tachometers and accelerometers are classical devices, but many types of position transducers are specially devised for use with numerically controlled machine tools.[6]

Some are directly connected to the leadscrews of moving components; this solution has as its main advantage great mechanical simplicity, but it does not readily lead to high accuracy, since the screw thread and the trueness of its bearing are liable to errors, the nut may have some backlash and the forces to which the screw is subjected introduce a torsional strain which is an additional cause of inaccuracy.

These difficulties have been more or less overcome by compensating for the error of the screws, as has been done for more than half a century in jig boring and scale-dividing machines; moreover, care is taken to connect the transducer to the end of the screw that is not subjected to torsion.

Despite all these precautions, the incorporation of control systems in the kinematic drive chain of the position transducer is only justified for machines of average precision, unless the mechanisms in question are small and subject to very small forces.

To improve the precision of measurements, transducers have been made independent from the drive system. Some are rotary transducers linked by a pinion to a precision rack. To be free from error, this system must fulfil at least three conditions: precision of the teeth, imperviousness to dust and lack of backlash in the gearing. The weakness of the transducer's driving forces ensures that there is no risk of tooth wear.

For a long time the manufacturers of jig boring machines have made use of a graduated scale mounted on a component fixed to the machine[7] and read by an interpolator system attached to a moving part. In another form, but based on the same principle, linear transducers have been made whose two basic parts—scale and reader—are attached directly to the components whose relative displacement is to be measured.

It is always very advantageous to make the origin of the machine coordinate system and that of the coordinate system of the part to be machined coincide. Otherwise the distance separating these two points must be measured after setting the part, and all the coordinates in the instructions must be corrected. These calculations cannot be carried out until the part has been set, causing an unavoidable delay.

The position the part must occupy on the table of the machine can also be defined in advance, but, in most cases, this demands very great care to effect this setting with a precision which is sometimes as fine as one-hundredth of a millimetre. Moreover, it would be necessary, in addition, to standardize rigorously the position of the origin in the work zone; this appears to be unrealizable, although highly desirable.

Of course, with pulse-counting measuring systems, whether they be absolute or relative (see p. 22), it is possible to set the origin anywhere in space if an arbitrary value can be placed in the counter when the moving parts are in any position. If the only value which can be set is zero, the origin must be situated within the surface accessed by the spindle.

If rotary transducers are used, it is possible to attach them to the drive mechanisms by means of clutches, and thus to set them to any desired angular position, but this method is hardly practicable where linear transducers are concerned.

This is why the most generally used solution is to fix an origin, generally at the end of the travel, and to introduce to all the measurements a correction whose value and sign are fixed when the workpiece is placed on the machine. This correction is input by a keyboard or decimal switches on the machine control console.

Types of transducers

In whatever way the transducers are attached to the mechanisms, the measurements are based on three principles:[8]

(a) Numerical measurement.
(b) Measurement by analogue comparison.
(c) Measurement by summation of pulses.

Many systems use a combination of two different methods, and sometimes three, one method being used for rough measurement, and the other for refinement.

These main systems are described in detail in the works on automation and machine construction, as well as in the specialized literature, and only a short description of their principles will be given here.[9–11]

NUMERICAL TRANSDUCERS

Numerical transducers supply directly numbers representing the linear or angular coordinates which they measure.

Numerical encoders

To read an encoder, various physical phenomena are used: the photoelectric effect (A.E.G., Heidenhain),[12] the Hall effect, pneumatic pressure drop,[13] etc. Systems of sliding electrical contacts have even been used (Harrison), but care must be taken during high-speed movements to raise the brushes which rub on the most rapidly moving contacts because failure to do this causes wear and reading errors.

Although the pure binary code[14] is used to a large extent, there are, nevertheless, many examples of the use of binary-coded decimal codes, cyclic codes, biquinary codes, Gray and Petherick codes, etc.[15–19]

All these codes have their own advantages, and for details the reader should consult the references.

As the computations are generally carried out in a pure binary code, the use of a similar transducer eliminates transcoding.

In going from one binary number to the next, several figures can change; for example, the binary number

1 0 0 1 0 1 1 1

is followed by the number

1 0 0 1 1 0 0 0

and one sees that the last four figures should change simultaneously. Now, the last four cells of the reader, whether they rely on mechanical, pneumatic, magnetic or optical phenomena, cannot change simultaneously, and three erroneous numbers are interspersed between the two correct numbers. For example, following the order of changeover, one might read successively:

	1 0 0 1 0 1 1 1	correct number
then	1 0 0 1 0 0 1 1	incorrect numbers
then	1 0 0 1 1 0 1 1	incorrect numbers
then	1 0 0 1 1 0 1 0	incorrect numbers
	1 0 0 1 1 0 0 0	correct number

The speed of electronic circuits is sufficiently great to register an incorrect number, however short it might be, to set in action inopportunely a logical process which can halt the normal working of the apparatus.

To overcome this difficulty, a method of reading based on the following observation is used: in a binary number, when the nth digit is 0, all the more significant digits are the same in this number and the succeeding number. If the digit is 1, there is identity between the more significant digits of the number and the preceding number.

On the tracks corresponding to the nth digit, where $n > 1$, two sensors are arranged to the left and right of the sensor on the first track, offset by an amount equal to one-quarter of the pitch of the preceding track.

According to whether a sensor on the nth track detects 0 or 1, the sensor to the right or the left on the $(n+1)$th track is selected. Thus, in Figure 2.3, we read cells 1, 2′, 3, 4′, 5 and 6. Because of the arrangement of the sensors, this is known as 'V-scan'.

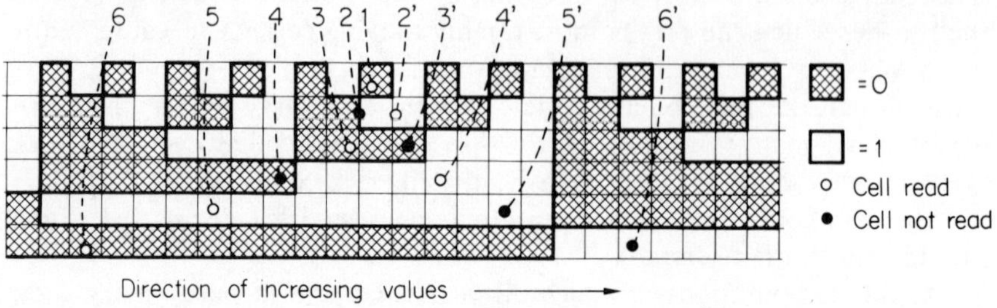

Fig. 2.3

If the gradations on the tracks are sufficiently precise, the sensors can be arranged in two parallel lines offset by one-quarter of the pitch to the left and right of the sensor on the first track (Figure 2.4). This is known as 'U-scan'.

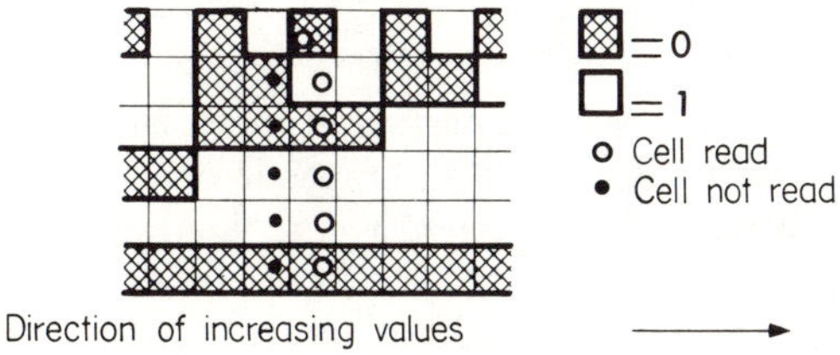

Fig. 2.4

The finest gradations which have been achieved to date have steps whose width does not exceed 5 μm. The optical reader covers several steps, and works through a grating of the same definition.

As the precision of a machine tool normally varies between 1 part in 10^5 and 0·5 parts in 10^5, the resolution of the reader must be at least 10^5 to 2×10^5 increments over the total travel. If a coded rule is used, there are, in principle, no problems, but if a circular transducer is required, it must be given a perimeter almost equal to the total travel, which is technically impossible for travels greater than a few tenths of a metre.

In such situations, another transducer (or several other transducers) is used to count the revolutions of the first. Each transducer is connected to the preceding one by a linkage whose ratio is equal to the number of segments on its least significant track (Figure 2.5).

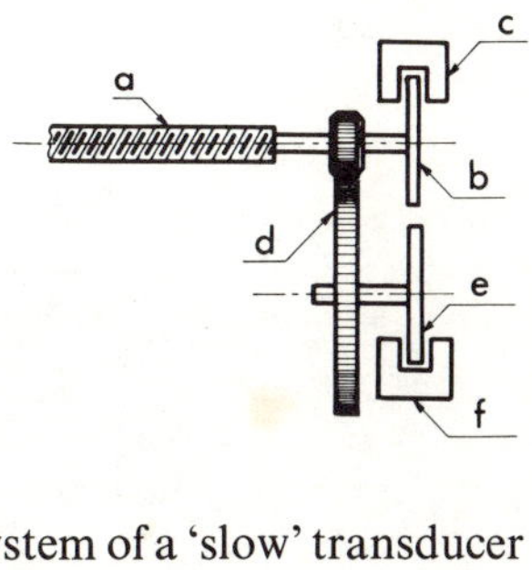

Fig. 2.5. a = screw with 5·12 mm pitch; b = 512 division coded disc (9 tracks), 1 division = 10 μm; c = reader for disc b; d = gear train, ratio 1:256; e = 256 division coded disc; f = reader for disc e

It would be impossible to control the reading system of a 'slow' transducer so that the change from one value to the next would coincide with the reader of the preceding 'fast' transducer changing from 1111 ··· 1 to 0000 ··· 0 or vice versa.

Two rows of readers in V or U configuration are arranged on the slow transducer; those on the least significant track are mutually offset by $n+\frac{1}{2}$ steps.

According to whether the reading of the most significant track of the rapid transducer gives a 0 or a 1, one or other of the rows of sensors on the slow transducer is read.

This arrangement is merely an extension of the V-scan or U-scan system.

Pulse generators

A pulse generator consists of a part, either translating or rotating, which emits a pulse, by means of an optical, magnetic or pneumatic system, each time it completes a basic displacement whose fixed magnitude is generally very small.[20–22]

The pulses emitted during a movement are summed in a counter. The total indicated is the displacement carried out since the counter was zeroed.

As with coded scales, several lines are read simultaneously if, for photoelectric or magnetic transducers, the resolution is less than 200 μm. With pneumatic transducers, a resolution down to 10 μm has been achieved.

As the form of the pulses is independent of the direction of displacement, the summation of pulses caused by a vibration of the moving part about the position where the signal changes must be avoided.

A 'direction discriminator' allows the counter to add or subtract automatically without having to determine by another means the direction in which the drive mechanism is revolving.

The principle on which discriminators are based consists in using two readers whose separation is equal to $n+\lambda/4$, where the length of a 0 or 1 segment is $\lambda/2$ and n is an integer (Figure 2.6).

If the displacement of the grating takes place in the direction F, the transducer A passes from 0 to 1 when B is reading 1 (Figure 2.6a) and from 1 to 0 when B is reading 0 (Figure 2.6b).

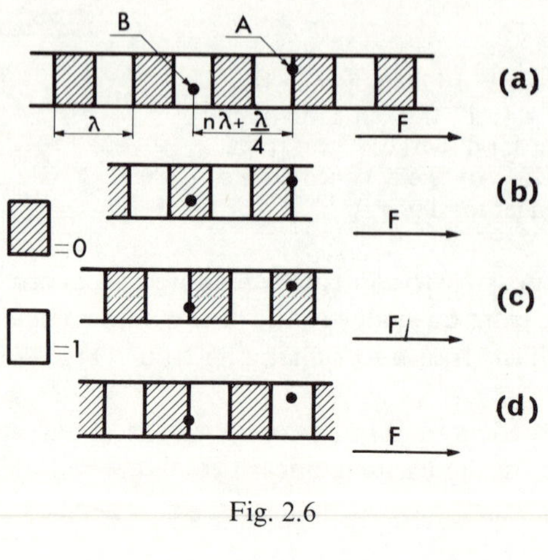

Fig. 2.6

If the displacement takes place in the same direction, B passes from 0 to 1 when A reads 0 (Figure 2.6c) and from 1 to 0 when A reads 1 (Figure 2.6d).

The results are reversed if the displacement of the grating is in the opposite direction.

The results are summarized in Figure 2.7.

	A *goes from*		B *goes from*	
	0 *to* 1	1 *to* 0	0 *to* 1	1 *to* 0
A = 0			+ →	− ←
A = 1			− ←	+ →
B = 0	− ←	+ →		
B = 1	+ →	− ←		

Fig. 2.7

In using this method, the resolution of the system has been doubled, because an impulse is generated each time it moves $\lambda/4$ instead of $\lambda/2$.

Precautions must be taken to avoid recording parasite pulses, generated by apparatus near to the control system, on the counter. The cables connecting the pulse emitters to the counters are shielded, and, if virtually absolute security is desired, the machine is fitted with an independent power supply isolated from the mains.

Other precautions are taken to avoid losing pulses. Firstly, pulse levels are kept well above the perception threshold.

Secondly, the machine is equipped with an auxiliary battery or independent power supply, to protect it from the consequences of interruptions in current which can pass unnoticed when they last only for a tiny fraction of a second, but during which no pulses would be generated.

Finally, to guard against the inconveniences of interruptions to the power supply, counters are fitted with memories.

When all these precautions are taken, the machines can operate for several weeks practically free from errors in computation.

However, to check that the work has been completed without error, after a certain number of movements, the occasional precaution is taken of returning the moving parts to a known position represented by a fixed stop. The value given by the counter is then checked against the expected value. If there is a discrepancy, this signifies that at least one error has occurred since the last check.

The advantage of pulse-counting systems lies in their simplicity, but, if all the precautions necessary to obtain a reliability comparable to that of systems with numerical or analogue transducers are taken, the prices become almost comparable, and the choice between the different methods depends only on their degree of compatibility with the remainder of the control systems.

ANALOGUE TRANSDUCERS

Conventionally, a quantity is said to be measured by an analogue method when its value is deduced from the magnitude of another quantity of a different kind which is related to it by a known relation of sufficient precision.

For example, these two quantities might be the resistance of a rheostat and the translation of its slider, or the displacement of a carriage and the angular position of its control wheel.

The phenomena used are extremely varied and the majority are mechanical, electrical or optical.[23]

Among the most standard examples are:

Mechanical phenomena: the rotation of a helicoidal surface causing the translation of a zero sensor.

Magnetic phenomena: the variation of an impedance as a function of the relative position of two parts of a magnetic circuit—Henry-Baudot–S.E.A. system,[24,25] Accupin system.

Electrostatic phenomena: the capacitance bridge—Reilly system,[26] Stop-Cote transducer (E.A.M.).

Electromagnetic phenomena: the phase angle of an inductive circuit—Inductosyn system (Farrand), Nultrax system (Westinghouse),[27] Muirhead,[28] rotary resolvers, etc.[29,30]

Photoelectric phenomena: measurement of the phase associated with the scanning of a scale carrying a grating—Ferranti.

The phenomena used for the reading and the measurement can differ; for example, a scale is read by means of an optical system. Null methods are generally based on the equilibrium of a bridge system, e.g. the Electrolimit impedance bridge (Pratt and Whitney), the photoelectric bridge.

With analogue transducers, the correspondence between the resolution and the total travel raises problems similar to those of numerical transducers, but as the current precision is of the order of 1 part in 10^3, improvement is only possible by utilizing extremely difficult and sophisticated techniques.

Most analogue transducers, therefore, have, side by side, several systems whose resolutions are in ratios of about 100:1 or 1000:1. Unlike rotary numerical transducers, no attempt is made to make the divisions of the component transducers coincide. When the reading given by one of them nears the desired input value, the next higher-resolution transducer takes over, and so on, until the highest-resolution transducer is used. This is the only transducer whose precision is a determining factor; the others only guide the whole system sufficiently close to its final position so that the final position is within a half period of the position corresponding to the input value.[29]

The Inductosyn linear analogue transducers show a similar arrangement; three gratings whose gradations are, respectively, 2 mm, 100 mm and 4 m are carried on the same scale.

MIXED TRANSDUCERS

From the preceding paragraphs, it will be seen that it is generally necessary to combine simple transducers. They need not be of the same type, and a great variety of combinations are used.[31–33].

In general, the primary standard consists of a divided scale or a counting system, and the interpolation is performed by an analogue transducer, but there are also systems in which the highest-resolution transducer is an encoder, and the coarsest stage is analogue, or works simply by counting the revolutions of the coded disc.

2.4 MODES OF OPERATION

There are many modes of operation, and several types of control system. Each has its own advantages: precision, speed, flexibility, reliability or price. They can be classified in terms of the work cycles they control.[34,35]

Point-to-point working

As was shown on p. 6, in this mode of operation the movements of the machine parts are independent and, in principle, only the final point matters for each part. Attempts to reduce the transit times have led to the creation of different systems whose sophistication varies according to the speed they achieve.

'YES OR NO' OPERATION

In this system, the comparison between the input value and the output value is qualitative rather than quantitative, i.e. the system only knows whether the difference is positive, zero or negative. If it is non-zero, its sign determines the direction of the displacement; when it is zero, the power supplied to the motor is cut off.

To avoid loss of time, a displacement speed of the order of 50–200 mm/s is chosen; so that the moving part has a degree of overshoot, which takes it beyond the assigned value. The comparator then receives a signal whose sign is opposite to the previous signal.

The movement is reversed, but the velocity is then reduced to around 5–20 mm/s, and the subsequent overshoot is smaller than the preceding one.

Once more the process is reversed, but with a sufficiently low velocity (0·05–0·5 mm/s) so that this time the overshoot is extremely small and lies within the required tolerance limits.

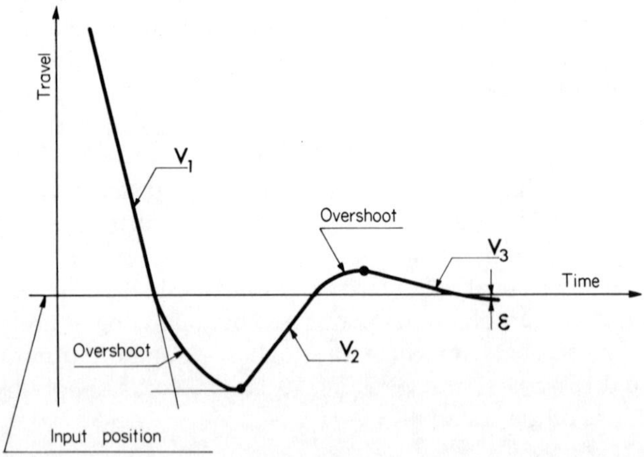

Fig. 2.8

This type of cycle is depicted in Figure 2.8.

To eliminate the effect of backlash, one or two movements of the part, depending on the direction of the initial displacement, are carried out at the intermediate velocity; so that the final movement is always in the same direction (Figure 2.9).

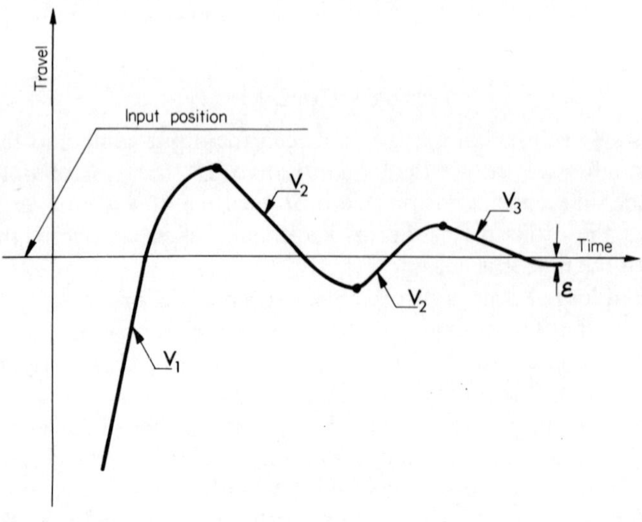

Fig. 2.9

VELOCITY-STEP OPERATION

As in the previous case, several velocity steps are used; the change is not made when the error signal is zero, but when it has attained a preassigned value. The comparator is quantitative rather than simply qualitative. The cycle is depicted in Figure 2.10.

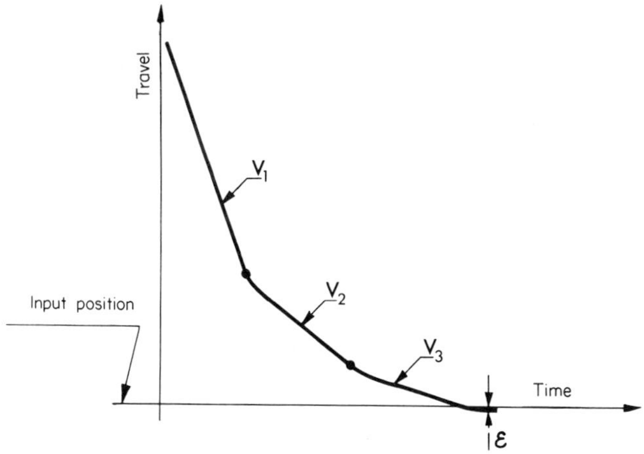

Fig. 2.10

As before, the direction of the initial displacement is taken into account in modifying the cycle, and in allowing a slight overshoot (Figure 2.11); so that the final movement at low speed is always in the same direction.

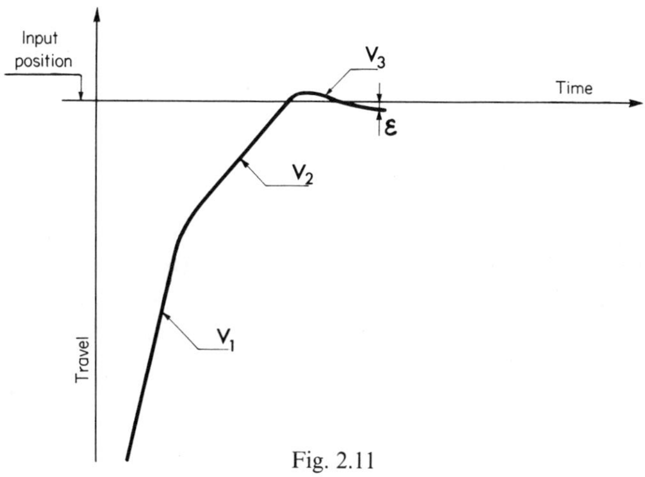

Fig. 2.11

Pulse-counting systems are particularly suitable for this mode of operation. There are two principal types. The simplest consists of an adder, which sums the pulses emitted during a displacement. If the counter is zeroed when the moving part is at its starting position, its position is measured continuously by the number in the counter.

In practice, a slightly different system is used. Instead of zero, the difference between the maximum number that the counter can hold and the number characterizing the displacement, to which is added a 'one', is placed in the counter. Thus, when the motion is completed, the counter will have returned to zero and will be ready to accept the instructions for the following movement.

This system is simple and has the advantage that, when a moving part passes at a given distance from the desired position, the counter always indicates the same value, simplifying the logic. If the capacity of the counter is expressed in the binary system by the number 1 1 1 1 1 1 1 1 1, the velocity-reduction point 16 units before the final position has the coordinates 1 1 1 1 1 1 0 0 0 0.

For a counter with a capacity of 2^{16} units, expressed in decimal by 6 5 5 3 5, and in binary by 111 111 111 111 111, the effects of these operations are tabulated in Figure 2.12.

	Value written in the counter	
Operation	*Decimal expression*	*Binary expression*
Zeroing of counter	0	0000000000000000
Entering of a displacement of 135·45 mm or 13,545 'units of 10 μm'	13,545	0110100111101001
Complementing	51,990	1001011000010110
Addition of 'one'	51,991	1001011000010111
Addition of 13,545 'ones'	65,536	1 * 0000000000000000

Fig. 2.12. As the 1* exceeds the capacity of the counter, only the number 0000000000000000 is registered

In this system, as each displacement is defined with respect to the preceding end position, the coordinates are said to be relative. This complicates the programming a little, and more so the reading of the position of a point with respect to the origin of the workpiece. Moreover, since the counter is simply a summer, the registering of parasite pulses arising, for example, from a vibration of the moving part which oscillates the reader segments, must be avoided.

Another system makes use of two counters. One is an adder and subtractor such that the number it contains represents the displacement of a moving part with respect to a fixed origin, which was the position occupied when the counter was zeroed. The second counter contains the number representing the absolute coordinate of the final position of the movement, and the commands are given according to the algebraic difference between the numbers in the two counters.

In this system, the coordinates of the velocity-change points are obtained by subtraction, causing a slight complication of the system logic.

To determine whether a pulse is to cause the addition or subtraction of a 'one', a discriminator is used (see p. 16).

In the single-counter system, the sign of the relative coordinate indicates the direction of the corresponding displacement. Moreover, if a moving part slightly overshoots the final value, a final position correction is made.

However, if this difference is less than the tolerance, error compensation can be dispensed with, provided that the summation of several errors is avoided. Thus this value is stored in the counter and, according to whether the two consecutive movements are in the same direction or in opposite directions, it is subtracted from, or added to, the previous coordinate.

CONTROLLED-VELOCITY OPERATION

If the positioning time is to be reduced to increase the output of the machines, the velocity is controlled by the error signal as soon as this has become less than a limit which corresponds to the optimal deceleration distance (Figure 2.13).

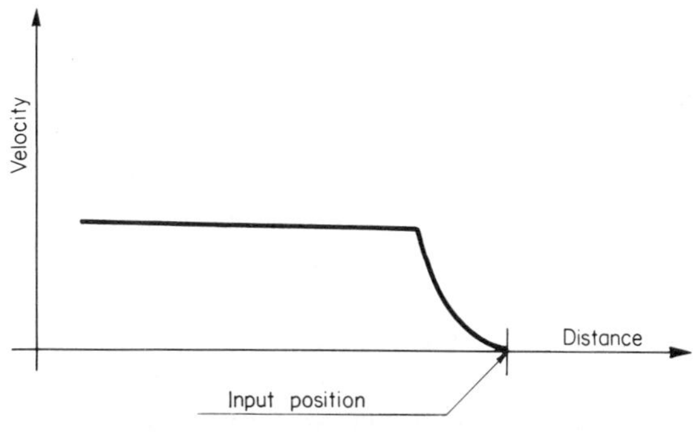

Fig. 2.13

PULSE OPERATION

In open-loop control systems, instructions are supplied in the form of pulse trains, which drive stepping motors.

These have the basic characteristic of rotating a small fraction of a revolution each time the control system receives an electrical impulse (see p. 35).

POSITIONING BY BLOCKS

Some systems (Schwarzkopf, G.S.P.) are based on the use of sets of additive blocks, this principle having been used for a long time to control the displacements of certain jig boring machines or of moving-table drilling machines with cross slides.

Generally they have drums carrying a set of ten blocks which differ mutually by 10 μm, 100 μm, 1 mm, etc.

To place blocks corresponding to the desired dimension between the stops, each drum is turned until the various blocks between the faces of the stops make up a stack whose length is, within a fixed limit, equal to the desired value.

In general, this system takes into account only the least-significant figures of the coordinates, and the system is completed by a precision rack in which the bolt of the measuring block engages.

The precision of this measuring system is of the order of one-hundredth of a millimetre, but it is obvious that imperviousness to dust is of prime importance, particularly as the presence of a foreign body can only be detected by observation, *a posteriori*, of systematic dimension errors in a finished piece.

Special apparatus is necessary to enable the origin of the measurement scales to be reset.

Straight-line milling

ADJUSTMENT OF COORDINATES

When a plane perpendicular to the axis of the spindle is milled, the cutting is performed by the front face of the tool, and it is only necessary to take into account, once and for all, the length of the tool in order to offset the origin of the corresponding scale.

When surfaces parallel to the axis of the spindle are machined, the tool cuts with its entire circumference, and the path of a point on its axis must be a straight line parallel to the directrix of the surface being cut.

The radius of the milling cutter must be added to, or subtracted from, the coordinates defining the directrices to obtain the coordinates of the path of the axis (Figure 2.14).

This correction can be made during the work preparation, but this necessitates fixing the cutter diameter in advance, and the use of two cutters of differing diameters to perform the roughing and finishing cuts consecutively using the same tape.

Other control systems permit the correction corresponding to the true diameter of the cutter in use to be entered. This data is entered by means of decimal switches on the control console. This flexibility is, of course,

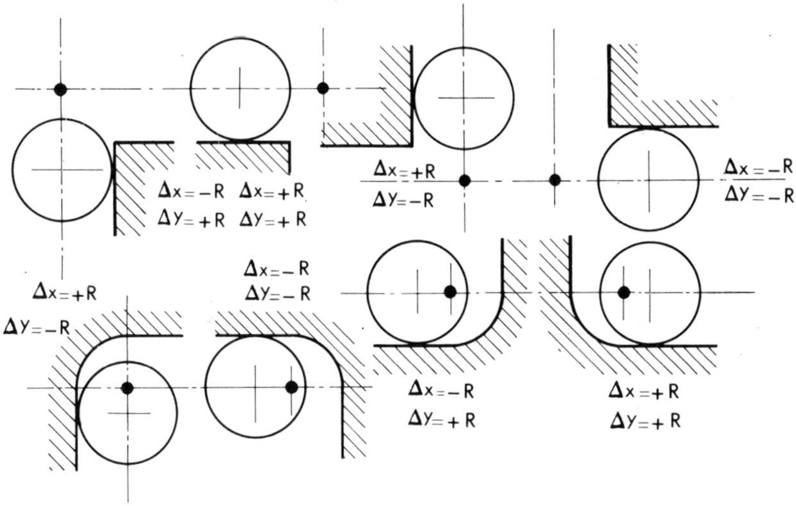

Fig. 2.14

advantageous, but brings a corresponding increase in the price of the control system.

OPERATION

Straight-line milling can be thought of as a particular case of point-to-point working. The control systems are thus rather similar, provided that the end of a cutting path includes, if necessary, a period of deceleration to avoid all risk of overshooting the end position. This naturally excludes the use of an additive blocks system, where the approach is always in the same direction.

Contouring

BASIC PRINCIPLE

During contouring, the speeds of the various moving parts must be related by a rigorously controlled value which is, moreover, subject to continual variation.

If the trajectory lies in a plane parallel to two of the machine axes, the contouring is said to be two dimensional. In the other cases, it is said to be three dimensional, even if the trajectory is a planar curve in a plane which is oblique with respect to the machine axes. The term signifies simply that three movements of the machine, at least, are liable to be used simultaneously.

TYPES OF CONTROL

Whatever modes of definition and control are used, the relative displacement of a tool with respect to a workpiece can be considered to be broken into elementary segments such as AB in Figure 2.15.

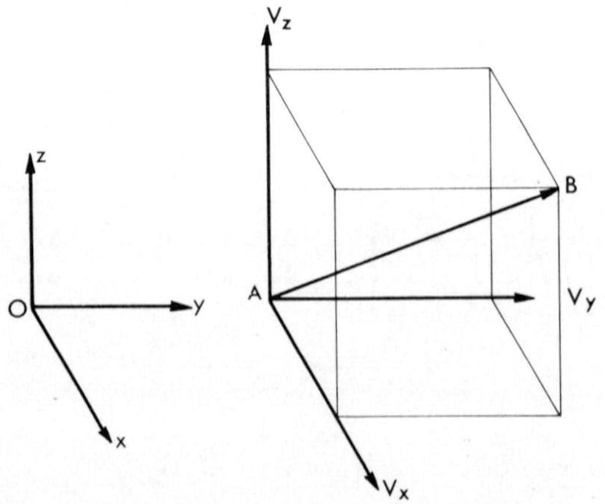

Fig. 2.15

If the initial position of the moving part is at A, and its objective is the point B, the displacement velocities imposed on the different parts by the control system are, respectively, proportional to the projections of AB on the reference axes.

If a constant relative velocity is required, the velocities must satisfy a further relationship:

$$V_x^2 + V_y^2 + V_z^2 = \text{constant}$$

Sometimes, to simplify the control system, only the largest of the three velocities is given a constant value; so that the resultant velocity can vary in the ratio $1:\sqrt{3}$, which is often admissible in profiling, where there is considerable freedom in the choice of optimal feed speed.

Some control systems implement the absolute proportionality of velocities, and the trajectory terminates, without any possibility of error, on the assigned end position. In contrast, with others the velocities are only approximated, and the resulting displacement is liable to deviate from the expected trajectory.

To compensate for this drift, which, by a process of accumulation, would lead to unacceptable errors, the input velocities are corrected at intervals of a few hundredths of a second.

When measurement of the coordinates shows that the moving part is, for example, at the position M, which is outside the segment AB (Figure 2.16),

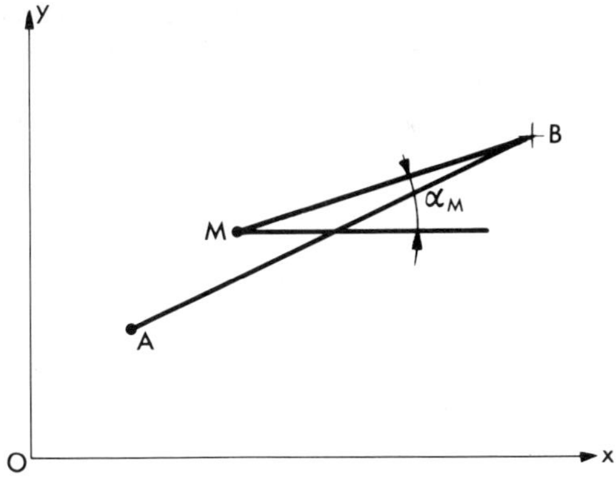

Fig. 2.16

one of two methods of correction may be selected. The first method is to compute the slope of the segment MB, and to reset the input velocities so that

$$V_y/V_x = \tan \alpha_M$$

At any instant, the trajectory takes B for its final value, and the method is known as pursuit control.

The second method aims at bringing the current point onto the ideal trajectory AB; this system is called trajectory control.

Taking into account the direction of V_M and the possibilities of accelerating the different parts, an attempt is made to give the axis a trajectory such as MN (Figure 2.17).

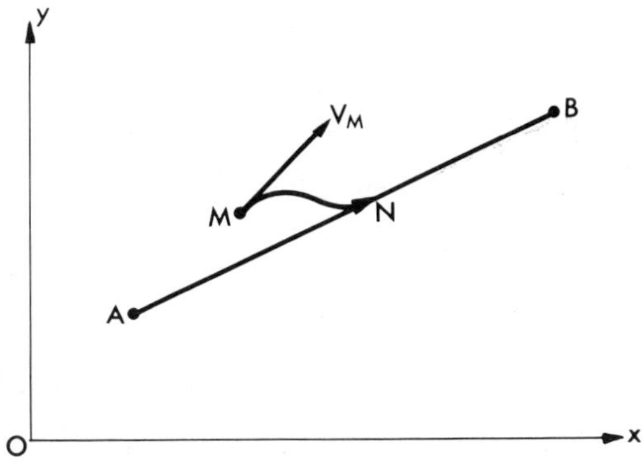

Fig. 2.17

If the characteristics of the control system are not adequately calculated, the return to the trajectory is too slow (Figure 2.18) or too fast (Figure 2.19), and the real trajectory then oscillates around the ideal trajectory.

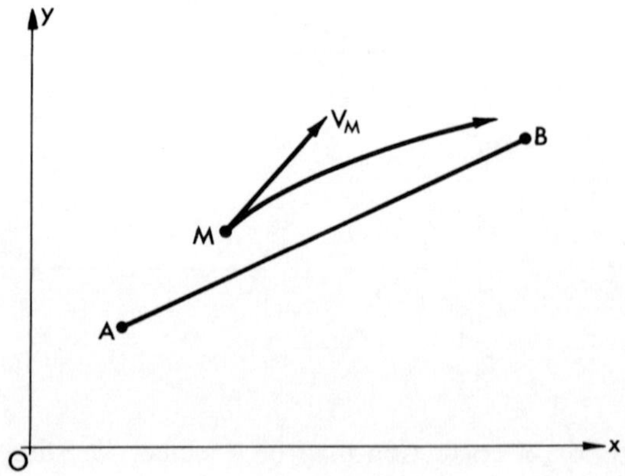

Fig. 2.18

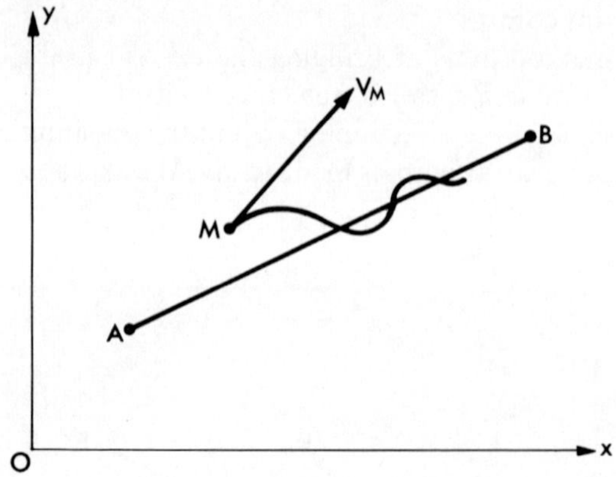

Fig. 2.19

Finally, in simple cases, correction can be dispensed with during the travel. The coordinates of the point are reset when one of them becomes equal to the required value.

If this point is B′ (Figure 2.20), the input values corresponding to the segment B′C are calculated. If the length of the segment is sufficiently small, distances such as BB′ lie within acceptable limits if the error in the angle α does not exceed, for example, 0·5°.

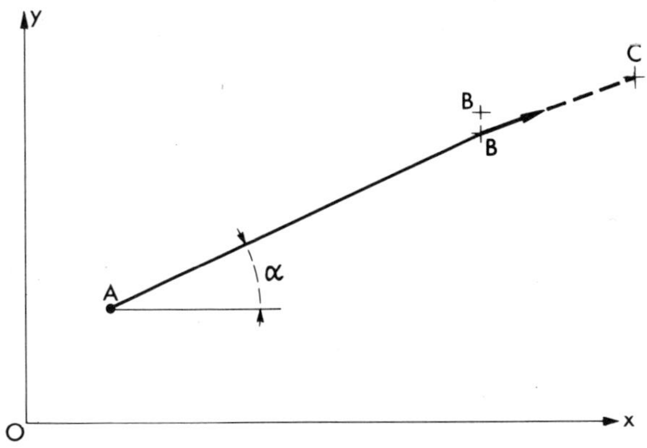

Fig. 2.20

METHODS OF DEFINITION

When any trajectory has been reduced to a certain number of segments, the control system must be supplied with the coordinates of the vertices of the resultant polygon.

Between these points, the moving part follows a trajectory which, according to the performance of the control system, coincides more or less exactly with a straight-line segment.

Different methods have been devised for calculating the coordinates of the vertices of the polygon which is an approximation to the theoretical trajectory.

Closely spaced definition

The distance separating the points is of the order of about 10 μm, i.e. it is of the same order of magnitude as the maximum error signal allowed by the control system.

The very large amount of data which must be supplied to the control system is generated in a special computer and written on a magnetic tape.

In return, the control system of the machine is relatively simple; it is called a non-interpolator system.

Medium-spaced definition

If a control system different from the one briefly described above is used, rectilinear segments may be fairly long without risking errors greater than the permissible tolerance. As a result, the points defining the path are much fewer than those in the non-interpolator system, thus reducing the number of calculations to be performed by the computer and the length of input media (generally punched tape) to be handled.

Given a trajectory, the tolerance defines the size of a region within which the approximating polygonal line must lie. According to the way in which the tolerance is defined, the trajectory forms one or other side of the region, or even its axis (Figure 2.21).

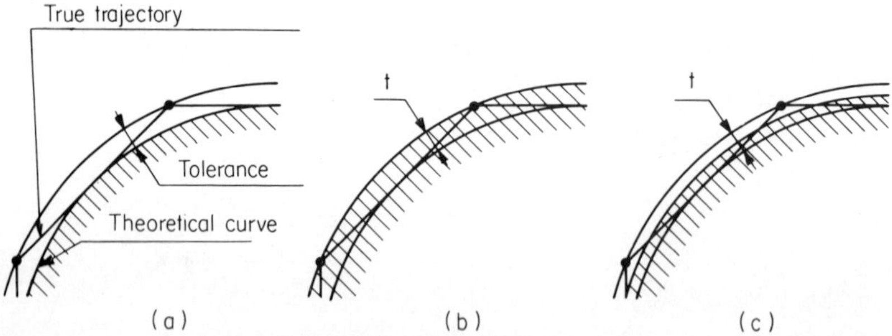

Fig. 2.21. a = outside tolerance, b = inside tolerance, c = straddling tolerance

If the radius of curvature of the trajectory is denoted by r, and the tolerance by t, the length of a segment of the polygon and the angle which it subtends are given approximately by the equations

$$l = 2{\cdot}8\sqrt{rt}$$

$$\alpha = 2{\cdot}8\sqrt{(t/r)}$$

The quantities l, r and t are expressed in the same units; α is in radians.

The results are processed to take account of errors in control.

The control system is called a linear interpolator, the adjective linear being used in the sense of rectilinear.

Widely spaced definition

To reduce the calculations to be performed by computer, and the instructions to be punched on the input media, curves which, in general, are arcs of circles or parabolic segments are inscribed instead of straight-line segments in the area defined by the theoretical trajectory.

On receiving from a punched tape the coordinates of the principal points A, B, C, D, E, etc., the machine's own control system works out the coordinates of secondary points situated on the circular arcs ABC, BCD, CDE, etc. (Figure 2.22), retaining only those which lie on the arcs AB, BC, CD, etc.

These arcs are not tangential to each other, but the program is designed so that the discontinuities are acceptable, taking into account the tolerances imposed.

If parabolic arcs are used, the number of points is increased from three to four, but, by automatic checking, it is verified that the parabolas obtained are neither degenerate nor imaginary.

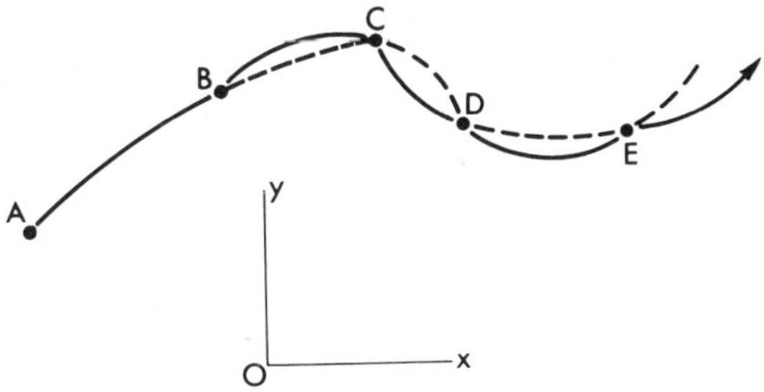

Fig. 2.22

Trajectories defined by segments of straight lines and circular arcs must be specially examined, as this is the most common case in mechanical parts, which are traditionally drawn using a straight edge and a compass.

If the control system also generates the coordinates of points on circular arcs, the information is reduced to its minimum. Some systems use the coordinates of the two ends and the centre of each arc, as well as the direction (clockwise or anticlockwise) in which it is traced.

Of course, these data are more than sufficient, and the calculation must check the compatibility.

The control system performs two distinct tasks:

(a) It calculates the coordinates of additional points on a curve whose parameters have been supplied.
(b) It guides the moving part from one point to the next along a trajectory which is approximately rectilinear.

Task (a) corresponds to the rôle of a computer or a function generator, and task (b) corresponds to that of an interpolator, in the proper sense of the word.

There are thus three principal types of control system:

1. Systems without interpolators working with an error signal corresponding to a lag a little greater than the distance separating two successive points.
2. Interpolators which impose the control strategy of trajectory or pursuit on the straight-line segments composing a polygon, which is inscribed in the tolerance band defined in relation to the theoretical trajectory, the vertex coordinates of the polygon having previously been computed.
3. The generator–interpolators which themselves calculate the coordinates of points between which they ensure rectilinear interpolation,

It is regrettable that the current terminology uses the same name for the last two categories, because this can lead to unfortunate confusion between types of apparatus whose potential and performance are very different.

General definition

If the majority of curves and surfaces currently used in mechanical engineering can be defined exclusively in terms of straight lines and circles, there are many fields where surfaces defined by other mathematical relationships need to be used. These, and surfaces defined by experience or intuition, must be translated into a form which can be used by an interpolator or a computer capable of handling only straight lines, circles or parabolas.

In general, these translation operations are performed using a computer. In certain applications, it is faster and more economical to use a generator capable of handling much more varied curves than circles and parabolas. However, one cannot try to translate the whole of any curve without first splitting it into segments whose form is sufficiently simple to be adapted to the mode of definition used by the chosen generator.

2.5 MOTORS

The reader will find detailed information on motors, their control and operating characteristics, by consulting the works on electrical technology, hydraulics and machine parts, and we shall confine ourselves here to considering their basic characteristics.

Linear motors

These are piston jacks driven by air or oil pressure. They are used to position a moving part in contact with a fixed stop defining an assigned position, and in this application their speed and their displacement are not subject to any control.

However, in the MOOG system, each piston is both a motor and a distributor, and controls systems working in both paraxial and point-to-point modes.[13]

Rotary motors

MOTORS WITH NON-CONTROLLABLE SPEED

These are asynchronous squirrel-cage motors; the speed variation (called slip) depends on the torque transmitted.

If their power does not exceed 50 kW, they are started simply by closing a three-pole relay without the need to use devices such as rotor resistances or stator connections. These motors can stand up to several hundred starts per hour.

According to the mode of operation for which they are intended, attempts are made to limit the starting current or to have a slowly falling slip–torque characteristic (Figure 2.23).

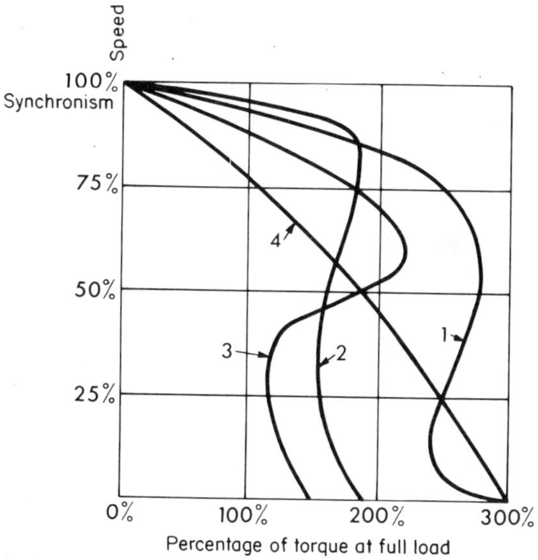

Fig. 2.23. 1 = weak starting current; strong slip. 2 = strong starting current; slight overload. 3 = weak slip. 4 = strong slip without stalling

Asynchronous motors produce no torque when they are unpowered. To absorb their kinetic energy rapidly, regulate their overshoot and provide a braking action, they are fitted with mechanical brakes released by electromagnets energized in parallel with the stator.

If the stator is double wound, the motor has several rotation speeds. Their ratios are generally 1:2 or 1:3, and sometimes 1:4. There are also triply wound motors, but they are rather cumbersome.

Multiply wound motors are used to attain high speeds, or to drive the slow-speed movements of machines working point to point and fitted with 'yes or no', or stepped velocity-control systems. The different speeds are obtained by using several motors linked by clutches, multipole motors, or mechanical gear changes.

SPEED-CONTROLLED MOTORS

These motors are widely used to drive contouring machines and for point-to-point machines whose velocity is controlled by the error signal.

The characteristics looked for in these motors are:

(a) Low rotor inertia.
(b) A large range of speeds in each direction of rotation.
(c) A stable speed which is little affected by the torque delivered.
(d) An appreciable torque at low speed and even, if possible, at rest.
(e) Good linearity.

Low rotational moment of inertia is a very important characteristic of motors with printed rotors (the S.E.A.–Henry-Baudot system).

Brush motors

Brush motors are driven by direct current or by rectified alternating current. All those which can be speed controlled have separate excitation.

If the ohmic voltage drop between the brushes and the transverse reaction of the armature are ignored, the rotational speed is proportional to the voltage at the brushes and inversely proportional to the stator flux:

$$n_{t/s} = k_1(U_V/\Phi_{Wb}).$$

On the other hand, the output torque is proportional to the inductor flux and the current through the armature:

$$C_{m.N} = k_2 \,.\, \Phi_{Wb} \,.\, I_A \,.$$

The operating speed is controlled by varying the voltage applied to the brushes or the exciting current. In the first case, the maximum output torque is almost constant, since it is only reduced by the voltage drop at brushes and the armature reaction. These are both approximately proportional to the rotational speed (Figures 2.24 and 2.25). The usable region is limited by the current which the armature can take.

When the speed is controlled by varying the flux, the power delivered is more or less constant. The torque, therefore, decreases rather quickly, and this method largely limits the speed ratio to 4:1.

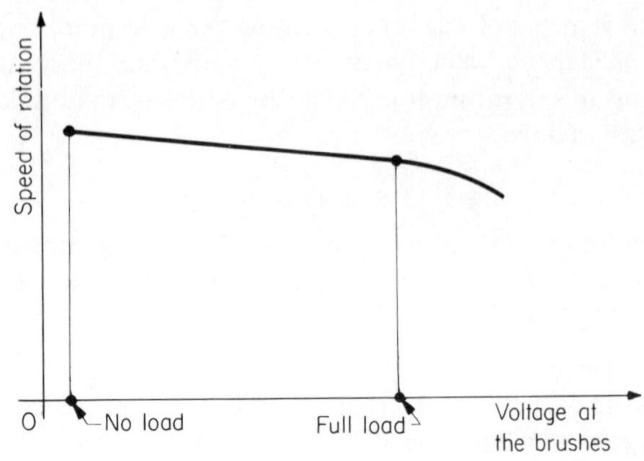

Fig. 2.24. Load characteristic with constant flux

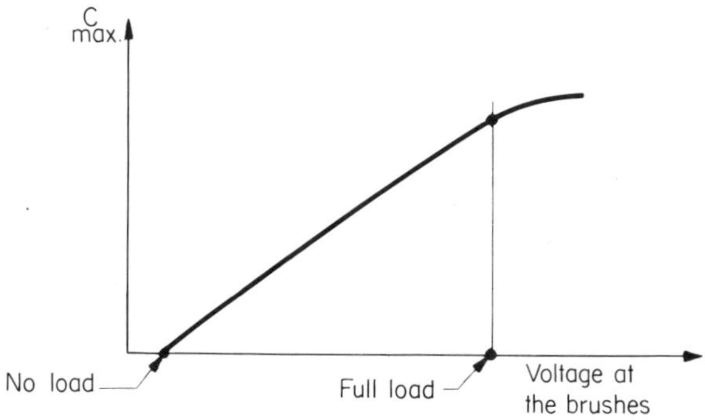

Fig. 2.25. Load characteristic with constant flux

Stepping motors

As we have seen on p. 23, stepping motors have been developed specially for open-loop systems using pulses.

The operation of these motors is based on a principle which has long been known, but their industrial use dates only from about 1955.

When a toothed stator and rotor are placed in a magnetic field generated by a winding (Figure 2.26), the rotor takes up a stable equilibrium position corresponding to the minimum reluctance of the magnetic circuit, i.e. the teeth align themselves face to face.[36–38]

If n assemblies with tooth pitch p are mounted on the same shaft, but staggered by p/n, and if the stators are excited one after another, the shaft carrying the rotors advances each time by p/n (Figure 2.27).

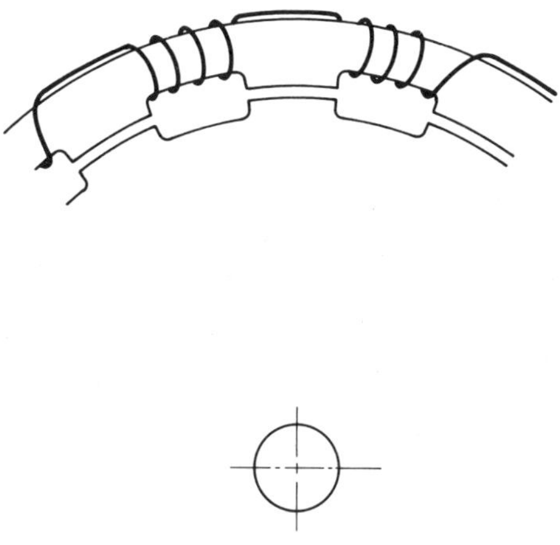

Fig. 2.26

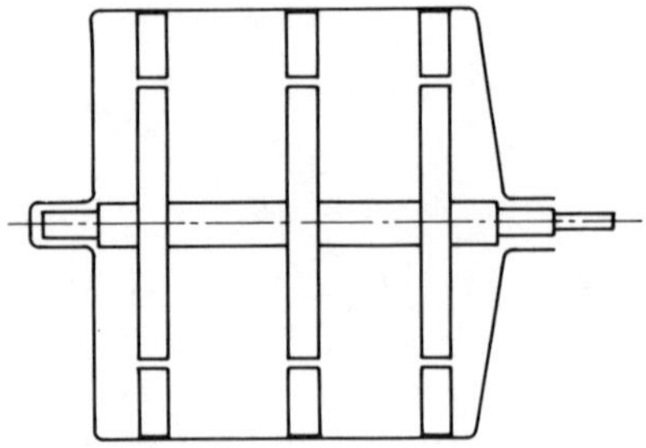

Fig. 2.27

A rotor generally consists of three to five wheels. If two neighbouring stators are powered simultaneously, the equilibrium position of the rotor lies between the corresponding equilibrium positions of the two stators, so that the number of possible positions of the rotor is doubled.

In general, motors are built with more than 200 positions per turn. Some have as many as 500.

The control system is activated by pulses. A static commutator successively activates the stators in the order 1, 1+2, 2, 2+3, etc. or in the opposite order, according to the instructions received.

To increase the torque, two or three rotors are activated instead of one or two, and the sequence is then 1+2, 1+2+3, 2+3, 2+3+4, 3+4, etc.

Although the rotor has a rather low polar moment of inertia, the torque it develops limits the angular acceleration it can produce without stalling. In the most favourable conditions, it will only start for a frequency less than half the maximum value it can accept, and provision is made for a starting scheme in which the frequency increases progressively.

The pulse frequencies that these motors can handle at present reaches 3 to 4 kHz, but certain high-performance models can handle 8 to 12 kHz.

The torque developed is restricted to low values, and motors of this type can drive directly only light mechanisms such as tape readers. If they are required to drive machine parts, they must be fitted with torque amplifiers.

Generally the amplifier is a hydraulic motor; the valve is rotary (Figure 2.28) or sliding (Figure 2.29). The latter system works with a lower input torque, but the former operates with a lower error signal.

The PIAT company builds stepping motors whose torque amplifier is an asynchronous motor. The frequency of the current which powers it is proportional to the frequency of the pulses received by the driving motor, taking account of the average slip in the amplifier.

As the speed of a stepping motor is proportional to the pulse frequency, they can be used in a tachymetric servo loop, but their most useful property

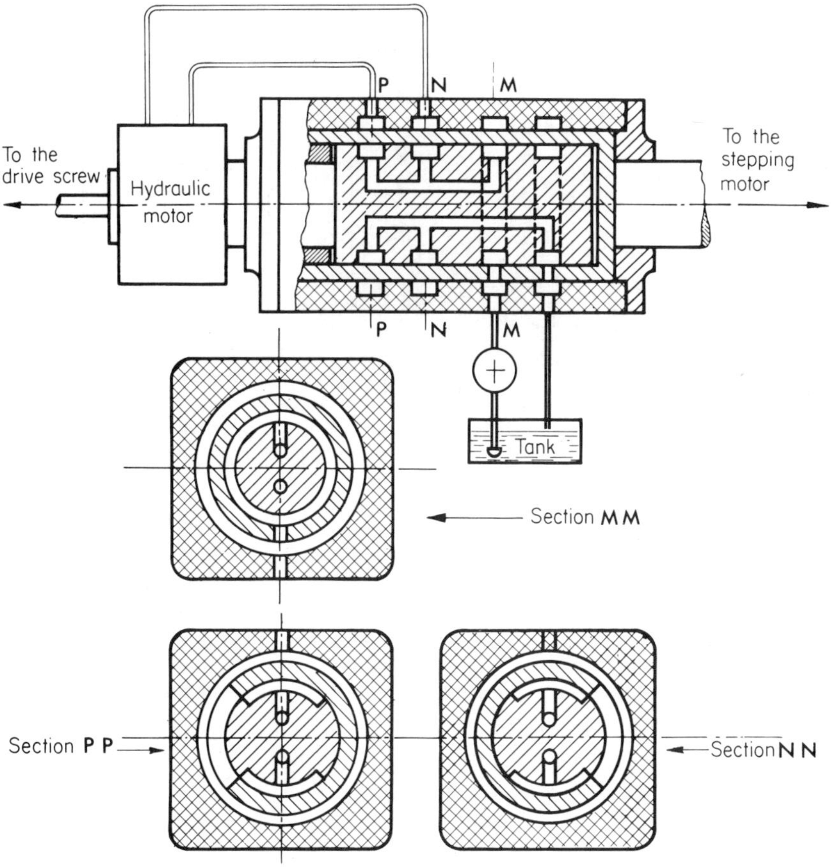

Fig. 2.28

is that of having a rotation angle which is rigorously proportional to the number of pulses received. As a result, stepping motors are capable of working in open loop, provided that they have been protected from loss of pulses, parasite pulses, or risk of stalling.

If a moving part has to follow a straight-line segment whose projections on the axes are Δx and Δy, and if each pulse causes a displacement d, each motor must receive, respectively, N_x and N_y pulses, these numbers being equal to $\Delta x/d$ and $\Delta y/d$.

Two counters, preset to the values N_y and N_x, are driven at the same frequency, and, each time one of them reaches its preset value, it returns to zero, emitting a pulse to its motor.

Thus, when the system has undergone $N_x \times N_y$ pulses, the two motors have received, respectively, N_x and N_y pulses.

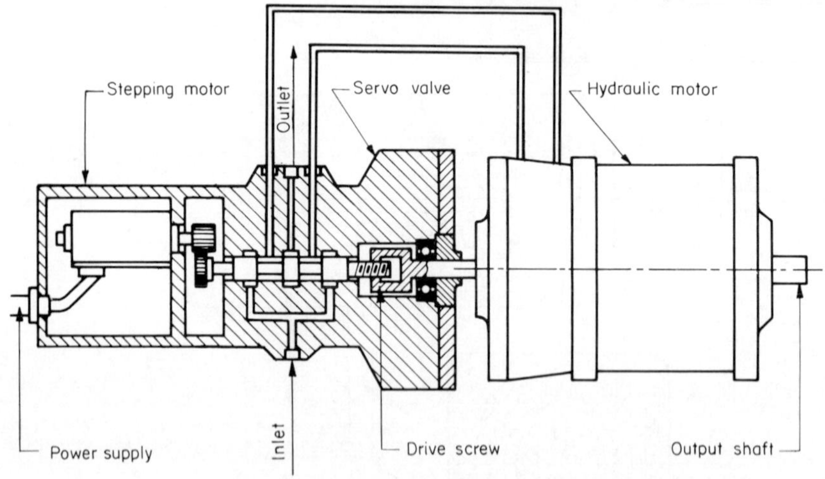

Fig. 2.29

If the length of the segment is

$$L = \sqrt{\Delta x^2 + \Delta y^2}$$

the speed of travel is V mm/s, and the distance corresponding to a pulse is d mm, the frequency is given by

$$f = \frac{N_x \, . \, N_y}{T}.$$

As $T = L/V$,

$$f_{\mathrm{Hz}} = \frac{N_x \, . \, N_y \, . \, V_{\mathrm{mm/s}}}{L_{\mathrm{mm}}}$$

we have

$$N_x = \frac{\Delta x_{\mathrm{mm}}}{d_{\mathrm{mm}}} \quad \text{and} \quad N_y = \frac{\Delta y_{\mathrm{mm}}}{d_{\mathrm{mm}}}.$$

Therefore

$$f_{\mathrm{Hz}} = \frac{\Delta x_{\mathrm{mm}} \times \Delta y_{\mathrm{mm}}}{\sqrt{(\Delta x_{\mathrm{mm}})^2 + (\Delta y_{\mathrm{mm}})^2}} \times \frac{V_{\mathrm{mm/s}}}{(d_{\mathrm{mm}})^2}.$$

If an approximation is sufficient, an expression which is more readily computed is used:

$$f_{\mathrm{Hz}} \simeq k \, . \, \frac{\Delta x \, . \, \Delta y}{\Delta x + \Delta y}$$

with

$$k = \frac{V_{\text{mm/s}}}{(d_{\text{mm}})^2}.$$

To control a three-dimensional displacement, the counters are set respectively to the values $N_y \,.\, N_z$, $N_z \,.\, N_x$ and $N_x \,.\, N_y$, and the frequency f is given by the expression

$$f_{\text{Hz}} = k \,.\, \frac{\Delta x \,.\, \Delta y \,.\, \Delta z}{\sqrt{(\Delta x)^2 + (\Delta y)^2 + (\Delta z)^2}},$$

Δx, Δy and Δz are expressed in millimetres, and

$$k = \frac{V_{\text{mm/s}}}{(d_{\text{mm}})^3}.$$

Hydraulic motors

Rotary hydraulic motors have extremely favourable properties: a large torque almost independent of speed, great dynamic rigidity, a large speed range, a high power/weight ratio and low inertia.

It is not surprising, therefore, that there has been a great development of these motors over the last decade, as they are very suitable for driving contouring machines.

All these motors have constant-displacement pistons. The swash-plate arrangement (Figure 2.30) seems more widespread than the radial arrangement (Figure 2.31).

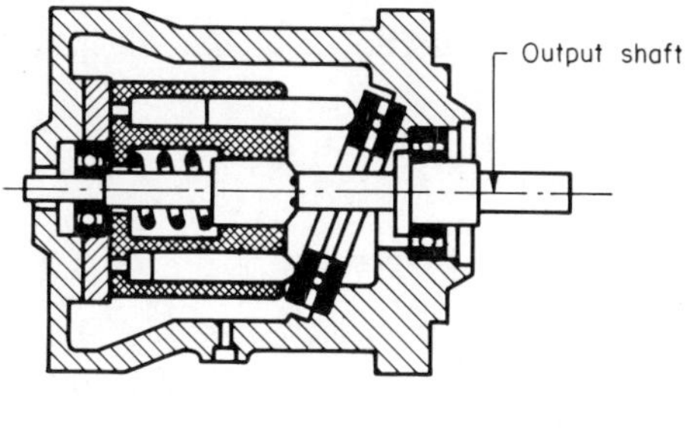

Fig. 2.30

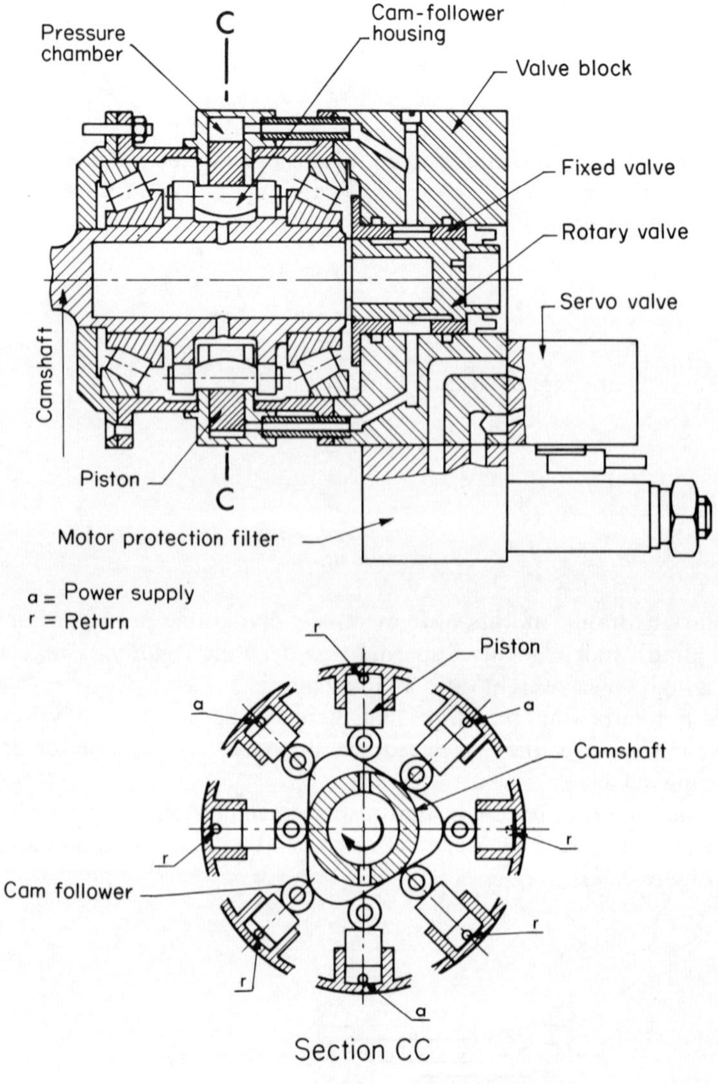

Fig. 2.31

References

1. *Standard for punched tapes* (*Norme des Bandes perforées*), NF Z 64 310.
2. *Standard for the ISO code* (*Norme du code ISO*), NF Z 62 010.
3. 'Digital magnetic tape: Will it win bulk of NC market?', *Steel*, 20 July 1964.
4. 'Magnetic tape controls attract new interest', *The Iron Age*, 25 March 1965.
5. 'Magnetic tape', *Steel*, 10 October 1966.

6. 'Numerical control systems, (Les systèmes de Commande Numérique)', *Bulletin du G.A.M.I.*, **1966**, special issue.
7. 'Measurement', *Product Engineering*, 29 March 1965.
8. 'Numerically controlled point to point positioning systems'; Part I, *Control Engineering*, January 1958; Part II, *Control Engineering*, February 1958; Part III, *Control Engineering*, March 1958.
9. H. Soubiès-Camy, 'The measurement of displacements by numerical methods (La mesure des déplacements par les méthodes numériques)', *Ingénieurs et Techniciens*, November 1966.
10. H. Soubiès-Camy, 'The measurement of displacements by analogue methods (La mesure des déplacements par les méthodes analogiques)', *Ingénieurs et Techniciens*, October 1966.
11. H. Soubiès-Camy, 'Analogue and numerical techniques ('Techniques analogiques et numériques)', Course at the École Supérieure d'Électricité.
12. F. J. Robinson, 'The use of encoders in machine tool control systems', *Machinery*, 1 January 1964.
13. A. W. Astrop, 'The MOOG hydra-point numerical control system', *Machinery and Production Engineer*, 2 February 1966.
14. 'The nature and scope of binary notation (La numération binaire et sa portée)', *Ingénieurs et Techniciens*, April 1965.
15. R. W. Hamming, 'Checking techniques for digital computers', *Control Engineering* May 1957.
16. I. S. Lerner, 'Digital computers need orderly number systems', *Control Engineering*, November 1955.
17. H. R. Glixon, 'Can you take advantage of the cyclic binary decimal code?', *Control Engineering*, March 1957.
18. F. H. Fowler Jr., 'Try permutation codes', *Control Engineering*, January 1964.
19. 'Harrison digitizers', Brochure of Harrison Reproduction Equipment Ltd., Farnborough, Hampshire, England (1964).
20. W. Seliger, 'A new optoelectronic positioning device (multiprimat) for machine tools [Un nouveau dispositif de positionnement optique-electronique pour machines-outils (multiprimat)]', *Microtechnic*, **XIX**, No. 3 and No. 6.
21. L. Rabian, '"Fotomat" numerical control for precision machine tools (Commande numérique "Fotomat" pour machines-outils de précision)', *Machine moderne*, December 1966.
22. G. Raimbault, 'Numerical control of positioning (La commande numérique de positionnement)', *La machine-outil*, June 1966.
23. J. Lombard, 'The measurement of displacements. The influence of numerical control on machine-tool design (Mesure des déplacements. Influence de la commande numérique sur la conception des machines-outils)', *CERMO*.
24. J. Henry-Baudot, 'A new linear magnetic measuring device for machine tools (Un nouveau lecteur linéaire magnétique de cotes pour machines-outils)', *Automatisme*, January 1958.
25. Raymond, Oulès and Pacory, 'Contribution to the control and automation of machine tools (Contribution à la commande et l'automatisation des machines-outils)', *Mécanique et Électricité*, May 1967.
26. 'Reilly linear transducer displacement transducer measuring system, *Machinery*, 12 April 1961.
27. 'Westinghouse Nultrax transducer for accurate linear measurement', *Machinery*, 15 March 1957.
28. 'Electric nut controls machine tool slides', *Metalworking Production*, 22 November 1967.

29. M. Eberle and G. Weibel, 'An absolute analogue system for position measurement for the numerical control of machine tools', *Siemens Revue*, February 1965, No. 2.
30. 'The Siemens sinumerik numerical control system', *Machinery and Production Engineering*, 14 September 1966.
31. A. T. Shepherd, 'Developments in Moiré Fringe measurement systems', *Machinery*, 20 December 1961.
32. B. J. Davies, 'A continuous path machine tool control system using coarse optical gratings (Staveley)', *Machine Tool Design Research*, **2**, Pergamon Press, 1962.
33. 'Polygon mirror eases NC fitting tolerances', *Metalworking Production*, 22 February 1967.
34. J. Lombard, 'A survey of the numerical control of machine tools (Aperçus sur la commande numérique des machines-outils)', *Compte rendu de la C.I.R.P. European Conference*, 24 *and* 25 *April* 1968, G.A.M.I.
35. J. Lombard, 'Introduction to the numerical control of machine tools (Introduction à la commande numérique des machines-outils)', *Mécanique et Électricité*, June 1968.
36. Electrohydraulic pulse motor, Fujitsu Limited', *Communications and Electronics*, No. 1015, Kamikodanaka, Kawasaki, Japan.
37. Vasiliev, Prokofiev and Wainberger, 'Step motors with an active rotor', I.F.A.C. Congress, 1963.
38. J. Derichs, 'The construction and mode of operation of stepping motors (Aufbau und Wirkungsweise von Schrittmotoren)', *Industrie Anzeiger*, 29 July 1966.

Chapter 3

Field of Application

The practical area of application of a technique is determined, in the last resort, by the technical and financial advantages it brings.[1–9]

3.1 PROFITABILITY

Increase in hourly cost

A numerically controlled machine is more expensive than a conventional machine with the equivalent capacity; the cost of the control system is added to the cost of the machine itself. Besides, the machine must be more precise, as its operation is automatic, and the operator cannot compensate, by his professional skill, for the errors caused sooner or later by wear in the equipment.

The increase in working costs depends on the one hand on the complexity of the numerical control; a point-to-point system is simpler than the system that is capable of contouring as well.

On the other hand, this relative increase is a function of the price of the machine. The price of the system is related to the type of work done by the machine, and depends very little on its dimensions. The cost of numerical control is therefore proportionally much greater for a small machine than for a large one.

The increase in hourly cost of a numerically controlled machine compared with a conventional machine of equivalent capacity can thus vary between 30 and 200 per cent.

Different economies, more or less easy to evaluate, can compensate completely, or partially, for this increase in operational costs.[10]

Direct economies

MACHINING TIME

Numerical control, in its present form at least, does not improve the performance of a tool or, as a consequence, reduce the cutting time required to carry out a given task.

On the other hand, it reduces the down time needed by the machine to move at its fast traverse speed from one point to another and complete its positioning.

This economy of time is not the only economy which can be attributed to numerical control. In fact, the operators are relieved of certain tasks, such as reading the drawings of the part, finding on them the dimensions defining each elementary operation, choosing and preparing the tools, and checking the dimensions, which they are compelled to complete for conventional machines before commencing the machining proper.

Furthermore, numerically controlled machines are designed to reduce other preliminary operations such as setting up, clamping and roughing out parts, as well as the location of datum points.

Finally, because the machining cycle is automatic, the operator can clean the work area during the cycle.

In conclusion, it may be assumed that the proportion of cutting time to working time, which varies between 10 and 20 per cent for ordinary machines, reaches 75 to 80 per cent for numerically controlled machines if the work is properly planned. Thus a numerically controlled machine can replace four or five conventional machines.

Some numerically controlled machines are fitted with devices which can, in a few seconds, replace one tool by another, whereas this operation, performed manually, generally takes rather longer.

The reader may be reluctant to credit this economy to numerical control. Other machines such as lathes have similar properties, and there is nothing to prevent other types of machines being fitted with similar devices; but it is the price of numerically controlled machines which justifies these devices being fitted to drilling, milling and boring machines.

The reduction of dead time is a great advantage for machines such as drilling machines, where the machining time does not make up the major part of the working time. It is less significant for machines such as copy millers, on which one job can last several days.

TOOLING

To manufacture any high-precision part with any degree of accuracy, it is not sufficient to machine according to marking out. Before the advent of numerical control there was no choice but to use high-precision machines, boring mills and jig borers or drilling machines complete with fixtures equipped with bushings (Figure 3.1).

The second solution is only justified financially if the tooling cost is amortized over a certain minimum number of parts; according to circumstances, the number may vary between ten and one hundred.

It is, therefore, justifiable to credit numerical control with economy in

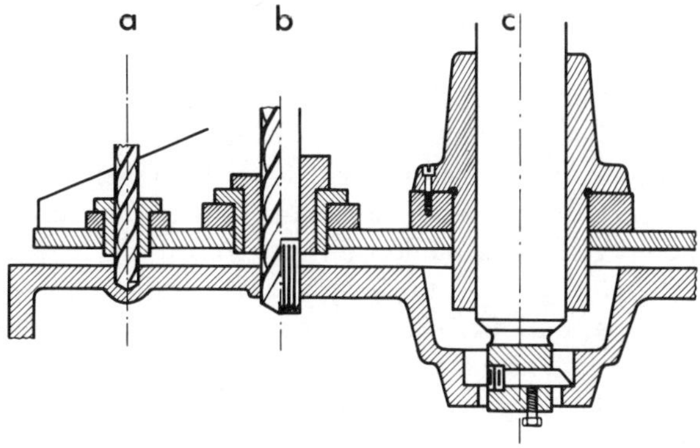

Fig. 3.1. a = simple drilling bush, b = interchangeable drilling and reaming bush, c = special boring bar bush

tooling costs, as well as the expenses connected with storage, maintenance, inspection, etc., of fixtures.

Some modifications frequently occur during the production of a series of parts when the production is spread out over some time. For this reason, the ability to modify the tooling is essential. As the first components have not necessarily disappeared from the catalogue, and may still be required, if only for spare parts, the tooling becomes multipurpose, and great care is necessary in identifying, for example, by colour coding, the bushings which are used for each type of component. This inevitably causes several risks of error. Numerical control allows the easy creation of distinct programs for similar, but nevertheless different, pieces, and this possibility is a definite advantage, but as difficult to evaluate *a priori* as a lapse of attention in the operator of a conventional machine.

Similarly, the elimination of marking out helps the operator; of course, marking out remains necessary where the correctness of the rough part needs checking before machining.

MANPOWER

When a machine works in an automatic cycle, it is possible, in principle, to limit the task of its operator to setting and removing the parts, handling the data input, placing tools in the tool magazine and cleaning the machine. While the machine is operating, he can prepare the next piece of work, examine the part, consult the drawings, familiarize himself with the instructions, check the condition of the tools, look for the clamps and jacks, or even set and clamp the next piece on a dummy table.

The practice of exploiting these capabilities is not yet completely widespread, but, as the reliability of systems improves, supervision by the operator will take up less and less of his time.

Of course, if punched tapes are not used, the operator remains totally employed in running the machine. He is responsible for inputting the instructions one by one, and if automation of the machine is restricted to positioning of the movable table, the quill is still hand operated.

Lastly, the operator of a numerically controlled machine does not need the theoretical and practical knowledge that is necessary to accomplish the same task on a conventional machine. As the work has been prepared by a methods department, he has neither calculations to perform, nor decisions to take regarding the sequence of operations or the conditions under which they are performed.

Nevertheless, the operator of a numerically controlled machine should be careful, attentive, calm and observant.

SHIFT WORKING

Because of the investment they represent, numerically controlled machines must, as a general rule, work two, if not three, shifts to be certain of profitable use.

As their operation is automatic, an operator can quite easily convey to his replacement the instructions and information necessary to complete the task in hand. This possibility exists with other machines, but there is no doubt that the greater degree of automation of numerically controlled machines confers an indisputable advantage which it is difficult to translate into figures with any degree of exactitude.

REJECTS

Of the rejects found in a factory, very few are attributable to machine failures, and human error remains their principal cause.

Thus it is not surprising that, once the reliability of systems became good, the rate of rejects from numerically controlled machines was much lower than that from other machines.

This is particularly noticeable in workshops where expensive and precise parts are manufactured, and this is sufficient to justify the use of numerical control for jig boring machines.

A corollary of this reliability is the reduction in inspection time. In general, only the first part in a series is inspected; the use of a punched tape guarantees that, save for a very improbable accident, no error can creep in during the machining of subsequent parts.

Thus inspection, which not only constitutes an additional expense, but also increases the time between two consecutive operations, is reduced very significantly.

Indirect economies

The use of numerically controlled machines sometimes has consequences in related fields. Some are technical, and others are administrative or commercial. The most important will be examined below.

JIG BORING

It is customary to rough out a part on a drilling machine and finish it on a jig boring machine.

Carrying out the two operations on the same machine without intermediate delay is rarely contemplated. The reasons for this mode of operation are evident:

(a) The use of a precision machine for roughing out is avoided; this would have the disadvantage of uselessly straining a delicate machine whose hourly costs are higher than those of an ordinary machine.
(b) During roughing out, a part generally undergoes a distortion or warping and certainly a rise in temperature. Between the two operations, correction is made, if necessary, of the starting points of the machining, and in particular of the resting face. Furthermore, the part is allowed to cool to the temperature of the jig boring shop (20°C).

If the part has been roughed out on a radial drilling machine according to the marking out, a large allowance must be left, amounting sometimes to 2 mm, so that the final machining will not risk running out of material. Under these conditions, there are often three, and sometimes four, boring passes.

In contrast, roughing out on a numerically controlled machine guarantees a concentricity of a few hundredths of a millimetre; so that it is sufficient to leave an allowance of from 0·2 to 0·3 mm between roughing and finishing, and to perform only one finishing pass.

COUNTERBORING

When parts have been radially drilled according to the marking out, the precision achieved does not guarantee the ability to assemble the bolts and screws in their housings, and, in general, a temporary assembly must be used to counterdrill the assembly holes.

This operation becomes costly once the pieces reach a certain weight, because they must then be moved by a travelling crane for assembly and disassembly, and it does not dispense with the final assembly after rough drilling and tapping on the radial machine.

Numerical control guarantees sufficient precision to eliminate the temporary assembly, and this allows the definitive assembly to proceed directly.

INTERCHANGEABILITY

Prototypes and items made in a short series by traditional means do not have the advantage of being interchangeable, unless they have been machined completely on jig boring machines.

As a result, replacement parts require counterdrilling and long and costly fitting, which is made superfluous by numerically controlled machines.

Even if the prototypes are never put on sale and rarely used outside the development and setting period, there are many products made in short series whose maintenance must be ensured after delivery by the manufacturer, who must also supply spare parts which can be fitted immediately with the minimum of adjustment.

HANDLING AND TRANSPORT

Numerically controlled machines are generally more versatile than conventional machines, since they can, for the most part, drill, tap, bore and mill.

Thus parts machined on numerically controlled machines pass between fewer machines than those made in the traditional manner. They are therefore subject to less transport, inspection and intermediate storage, which is an appreciable economy.

LEAD TIMES AND WORK IN HAND

The versatility of numerically controlled machines, as we have seen in the previous paragraph, reduces the number of operations in a range of work. The manufacture of a part is thus speeded up, and this advantage is most valuable when a prototype is being set up or a broken mechanism is being repaired.

Besides, the number of parts under construction is reduced, which reduces investment, since the parts represent nonproductive capital whose return constitutes a supplementary financial charge.

STOCKS

When some parts have already been made on numerically controlled machines, the production of a new series is fairly simple, since the different punched tapes that are preserved in the archives can be used to print the instructions and work tickets or to control the operation of the machines.

Hence it is possible to reduce the quantities produced in each run, and not to initiate a run until the stocks have reached a level lower than is normal in conventional manufacture. The average stock is thus reduced, lessening to the same degree the return on capital that it represents.

OVERHEADS

An appreciable part of the price of a factory is the cost of buildings, together with their heating and lighting systems, as well as their outbuildings, changing rooms and showers, etc.

Under the preceding headings, we have seen that the surface area of a numerically controlled machine workshop is smaller than that of a traditional workshop because of reductions in the number of machines, manpower, stocks, lead times, tooling shops, inspection stations and preliminary assembly working. However, the use of numerical control brings a certain expansion in the planning office.

ARCHIVES AND ADMINISTRATION WORK

The archives of a numerically controlled machine workshop can be reduced to a collection of punched tapes; as a final step, they could be transcribed onto magnetic tapes and accessed automatically, simplifying the administrative work to a great extent.

3.2 *TECHNICAL POSSIBILITIES*

Numerical control, when it appeared in 1942, had been developed to solve the problems of contour milling. It was next applied to drilling and reaming.

Apart from some enterprising attempts made by the manufacturers of systems and machines in several fields, the industrial applications of numerical control remained limited almost exclusively for twenty years to those two modes of operation. Contouring was more often two-dimensional than three-dimensional.

However, the control systems gradually acquired almost complete security of operation at the same time as their basic components became standardized.

When the eventual users had gradually gained confidence in a technique which had brought great changes to their ways of working, and when the price had become sufficiently low for the threshold of profitability to be reached comfortably, the applications of numerical control multiplied, sometimes in unexpected ways.

It is not possible to give a complete list, and only the best known applications will be mentioned here, with their main characteristics.

The characteristics arise out of two quite distinct techniques: point-to-point and straight-line positioning, and contouring. Not all the operations belong, in the strict sense of the word, to the field of machining, but numerical control is a technology whose subdivisions are all related, and one cannot intentionally omit some without being detrimental to the overall picture.

Point-to-point and straight-line operation

DRILLING, TAPPING, REAMING AND MILLING

These are the conventional machining operations. The principal types of machines will be described in Chapter 4.

PUNCHING

Conventional punching machines have been modified to work by numerical control: a moving table is attached (Figure 3.2) and the plate, held by two clamps, is positioned under the punch head.[11]

The punch is operated by a hydraulic or pneumatic jack. Numerical control also operates two rotating turrets on either side of the working plane. One carries the punches, differing in shape and size (circles, squares, slots, etc.); the other holds the corresponding dies.

A precision locking mechanism ensures the alignment of the die and the punch.

The operation of punching proper lasts only a fraction of a second, and it is essential that the positioning should be extremely rapid, as otherwise the output of the machine would be low. On the other hand, great precision of location is seldom required.

Speeds of 100 to 200 mm/s are not uncommon, the positioning time being a few tenths of a second.

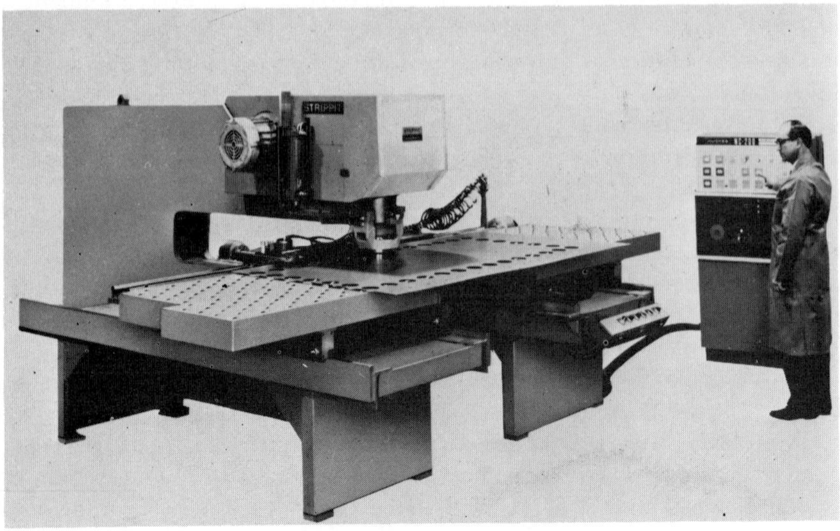

Fig. 3.2. Strippit punching machine

SHEARING

In high-production factories, the general tendency is to receive plate in the form of rolls.

Before being fed to the presses, the plate must be straightened and sheared to an appropriate size. As this operation is very brief, and its rhythm is much faster than that of stamping, the straightening and shearing machine often changes its program, and its setting is subject to frequent modification. Furthermore, it must be somewhat precise (within a fraction of a millimetre), as each positive error causes excessive consumption of material, and each negative error causes the rejection of a part.

The plate used is measured by the rotation of the straightening rollers, and these are stopped at the moment of shearing.

As slight errors in the angle of rotation do not have grave consequences, provided that they do not build up, a pulse-counting system is used.

RIVETING AND SPOT WELDING

These operations, as far as their procedure is concerned, are similar to punching. As with punching, the overall output depends greatly on the rapidity of positioning, and high-performance control systems are sought.

BENDING

Conventional tube- and plate-bending machines can be controlled by a numerical system[12,13] (Figure 3.3).

The simplest method is to count pulses corresponding to a travel determined by the chosen program.

However, bending of tubes and plates is an operation which depends on the physical properties of the material being worked. These are not known in advance, and are not constant from one end of a rod or coil to the other. After a piece has been subjected to its maximum deformation, it springs back, and, if this is not equal to the expected value, the bending angle is not held.

To overcome this inconvenience, a system has been developed, which, after measuring the mean error of a number of parts, applies a correction to the operation in hand, and this allows the maintenance of almost constant results, although the characteristics of the material vary from one part to another.

A similar technique could be applied for correcting parts distorted during machining or heat treatment, the force applied being calculated as a function of the distortion to be corrected. At present, the development of this process is proving extremely difficult.

Numerical control, whose advantages generally belong to small runs or one-off jobs, here finds a use in repetitive work. However, as bending operations are dependent on unforeseen physical data, they pose a unique problem

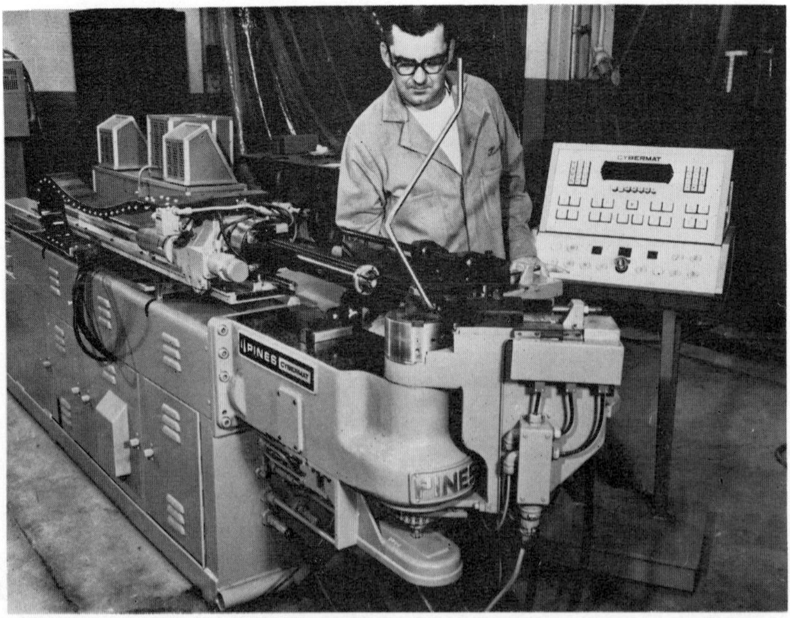

Fig. 3.3. Pines bending machine

in every case. They are really, therefore, one-off jobs repeated with great frequency, and it is not surprising that numerical control finds a special application.

WIRING AND ASSEMBLY

In the electronics industry, and especially in the computer industry, tasks are carried out which are minute and relatively seldom repeated. Numerical control plays an important role—it is applied to the wiring of connection panels and the mounting of components on printed-circuit boards.

Connection panels currently carry several hundreds of pins, sometimes more than a thousand, which are connected in pairs by wires.

Each wire is cut to the required length, the insulation is stripped at each end, and a special wrapping device makes the two connections after the wire has been placed in the selected path (Figure 3.4).[14,15]

Machines for the mounting of printed-circuit boards receive the components arranged on loaders or lined up in a strip. The place assigned to each part is defined by the coordinates of the holes locating the ends of their connecting wires. These holes are located at the intersection of a 0·1 in grid (Figure 3.5).[16]

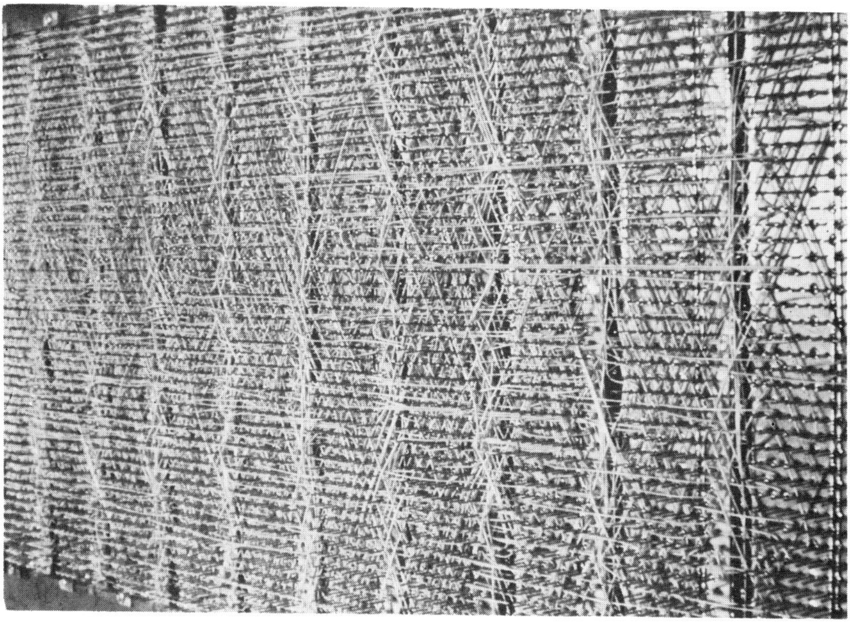

Fig. 3.4. Back-panel wiring

Fig. 3.5. I.B.M. circuit-mounting machine

BALANCING

In order that a rotating body should not subject its mountings to periodic forces, one of the principal axes of inertia must coincide with its axis of rotation.

The relative position of the two axes is determined by rotating the body to be balanced in its bearings; the forces transmitted to the bearings, or the amplitude of their displacements if they are mounted on elastic supports whose natural oscillation frequency is considerably less than that of the rotary movement, are measured.

When the characteristics of the central ellipsoid of inertia are known, the position and the weight of the masses to correct the unbalance are computed in two planes which are orthogonal to the axis of rotation.

This correction can be effected by adding weights, i.e. by welding a slug or strip of steel tape or else by removing metal with a milling cutter or drill.

The data resulting from the calculations is stored in a numerical or analogue form, and controls the orientation of the part and the depth of the cut.[17]

It would be extremely difficult to design a universal balancing machine, since it would be necessary to compute the characteristics of the inertia ellipsoid of each part. Therefore numerically controlled balancing machines are intended for mass production. A balancing machine, like bending and correcting machines, works with unforeseen data. Each operation requires the solution of a particular problem, and one could say in this situation that numerical control is applied to a single problem which is repeated with high frequency (Figure 3.6).

INSPECTION

The flexibility, speed, precision and reliability of numerical control suit it well to inspection.

The simplest machines check various properties: dimension, weight, surface finish, etc. The measurements, stored in numerical form, control the identification and classification of the inspected parts.

These machines are very specialized and are only used in mass-production runs for components such as ball and roller bearings, pistons and piston rods, shock-absorber valves, etc.

Others are used to check many dimensions of a prototype part, or of a part sampled from time to time from a series.

Some (Pratt and Whitney) are similar to column drilling machines, others (D.E.A., Heyligenstaedt, Ferranti, Aristo, etc.) take the form of a carriage carrying a slide fitted with a measuring probe (Figure 3.7).

Fig. 3.6. Renault crankshaft-balancing machine

Fig. 3.7. Heyligenstaedt measuring machine

For a certain number of points, two of whose coordinates are known, the machine measures the third, or the difference from its theoretical value, and delivers the result in printed form or on a punched tape.

Contouring

TURNING

Numerical control can direct the movements of a lathe, since the trajectory of a tool depends on the generatrix of the part being made. It is applied to spindle lathes, vertical lathes and chuckers.

The majority of elementary turning operations are very brief. They are of medium precision, since, if close tolerances are required, finishing is often performed on a grinding machine. The rough part is generally modestly priced, especially if it is sufficient to make it from rolled bar.

For all these reasons, a pulse-counting measuring system can be used in many cases, as it is more economical.[18]

FLAME CUTTING

Oxy-acetylene cutting apparatus has been fitted with numerical control (Figure 3.8). A correction in the direction of the normal to the trajectory allows for compensation of the error introduced if the size of the cut is not taken into account.[19]

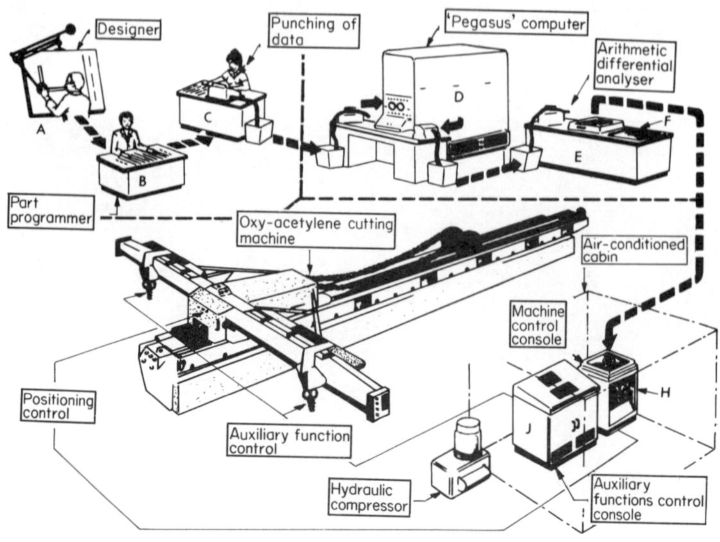

Fig. 3.8. British Oxygen/Ferranti oxy-acetylene cutting system

WELDING

The aerospace industry uses numerically controlled welding machines. The trajectory of the electrode head is generally a twisted curve, so a machine fitted with a follower and a three-dimensional template would be difficult to construct[20] (Figure 3.9).

Fig. 3.9. Sciaky welding machine

GRINDING

Numerical control is applied to various types of grinding machines.

Cylindrical grinding

Since about 1960, centre grinders have been exhibited (Landis, Fortuna, etc.) that have been fitted with numerical control for the longitudinal positioning of the table and the adjustment of the automatic measurement of the part.[21]

It seems that the price of the system, if it is compared with that of the set of dial gauges it replaces, makes its economical justification rather difficult, and at present this type of machine is not widely used.

Grinding of camshafts

In high-production factories, the grinding of cylindrical cams is performed by copying machines.

To make the master cams, the conventional method consists of machining a standard part by milling or grinding a number of facets defined at intervals of 1° or 0·5°, according to the cam's action. The edges are then smoothed down by hand using an abrasive stone.

For cams operating at high speeds, very precise positioning is looked for, but even greater attention is paid to the velocities and accelerations that the positioning produces, since each error is translated into shocks which cause hammering, abnormal wear or spring bounce.

Moreover, the movements of parts of the grinder require dynamic forces which produce distortions that are very difficult to foresee. Thus the manufacture of the master cam is accomplished by successive approximations.

The corrections to be applied are expressed in micrometres, and the difficulty of the task of specialists responsible for cam manufacture can be imagined.

This work requires minute care and great professional skill, and the time required for machining a part sometimes reaches several months, a serious manufacturing inconvenience.

Two Japanese companies (Toyo-Kogyo and Hitachi) (Figure 3.10) have built special numerically controlled grinding machines that have been developed in the Fujitsushinki laboratories and are based on the use of stepping motors driven by pulse counting.[22]

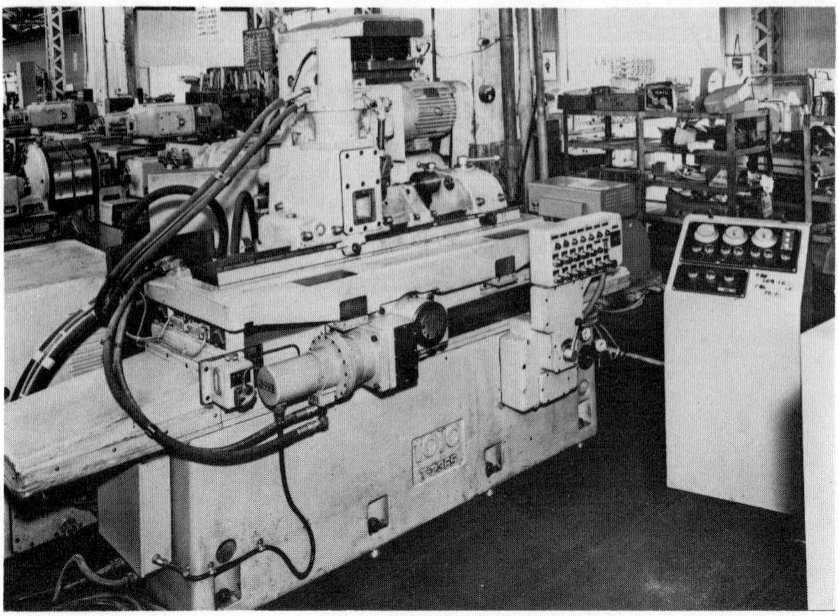

Fig. 3.10. Toyo camshaft-grinding machine

In this system, the time for preparing the control tape is less than 48 h, which greatly reduces the manufacturing time and also the manufacturing cost of the prototypes.

This process has been shown to be economical, not only for the manufacture of standards and prototypes, but also for small series and replacement parts, since its use dispenses with the making and storage of costly and bulky master cams.

Grinding of arbitrary surfaces

Gas turbines contain refractory-alloy blades and three-dimensional governor cams which must be hardened to resist localized friction forces which might produce excessive wear.

The American Ex-Cell-O Corporation has built milling machines and special grinding machines fitted with a sphero-conical tool to machine this type of part. Their control system is numerical, avoiding the manufacture of the master cams usually used for this type of work.

Some grinders, of the column and moving-table type, are also numerically controlled, but these machines seem not to exist in any great number at present.[23]

DRAFTING AND DESIGN

To make templates whose precision is expressed in tenths of a millimetre, one could simply cut them out roughly and finish them with a file following a scribed mark.

If this is to be applied to parts made in more or less lengthy series, such as the ribs, stiffeners and structural panels of aircraft assembled by riveting, the procedure is the same, but instead of scribing the outline directly, it is printed by means of a scribed plate acting as a negative.

Precision drawings are equally necessary in topography and cartography, as well as for the manufacture of the standards used on the projection screens of inspection machines, and grinders and oxy-acetylene cutters operated by the photoelectric scanning of an optical follower.

Numerical control has found a large field of application here, and numerous applications of American (Gerber, Universal Drafting Machines, Benson, Tridea), English (Cramic, Ferranti), German (Aristo), Swiss (Contraves) and other machines could be cited[24] (Figure 3.11).

These machines, some of which can reach a drawing speed of 300 mm/s, are also used for checking punched tapes intended for lathes or milling machines. The projections of the relative trajectory of a tool are thus quickly obtained, and the user can check that there is no large error in the program; otherwise long and costly dry runs on the machine tool itself would be necessary, carrying serious risks of an accident when the cutter path is in error.

Fig. 3.11. Gerber drafting machine

When the only task of the machines is to draw predefined curves, they can be considered as drafting machines. If they are fitted with a system that is capable, without appreciable delay, of producing curves, the data for which is defined by the operator on the spot, they truly become design machines.

The control system in this case plays the rôle of generator, interpolator and controller, as was indicated on p. 31.

Besides design machines proper, use is made of cathode-ray tubes to produce drawings, schematics and curves characterizing various phenomena. The dimensions of such apparatus seldom exceed 500 × 500 mm, and their precision is very much lower than is required in design work. Thus it is necessary to use 'tricks' to correct the distortion, and to adopt scale changes to overcome the lack of precision.[25]

WINDING

Containers made by winding steel wires or glass-fibre filaments impregnated with resin around collapsible mandrels have been used for about fifteen years for the storage of gas and liquids under pressure.[26]

The rotation of the mandrel and the translation of the wire guide are constantly coordinated, and numerical control is well suited to this type of work (Figure 3.12).

Fig. 3.12. Sud-Aviation winding machine

MILLING

Two-dimensional milling

Under this heading comes the operation of a machine whose tool axis undergoes a translation defined by a curve in a plane parallel to one of the faces of the trihedron of reference and normally perpendicular to the axis of the spindle itself. In general, the tool is cylindrical, conical or toroidal. This mode of operation is often designated by the symbol '2D'.

Stepped two-dimensional drilling

The surfaces which one wishes to generate are not always faces with constant taper.

In order not to increase unnecessarily the complexity of the program, they are divided into a series of elementary surfaces of constant draw, which have, for directrices, intersections of the volume to be machined with planes parallel to one of the faces of the axis system, and are, in general, perpendicular to the axis of the spindle.

This type of working is designated by the symbol '$2\frac{1}{2}$D' or '2D + 1'.

Three-dimensional milling

In the most general case, a surface is defined by twisted generatrices. To sweep out such trajectories, it is necessary to coordinate at least three movements.[27]

If, in addition, the orientation of the spindle axis is related at each point to the normal to the surface it is machining, the machine must have at least two rotational movements as well as the three orthogonal translation movements.

Whether a machine is equipped with three or five movements, its mode of working is designated by the symbol '3D'.

3.3 *ECONOMIC LIMITS*

In the preceding paragraphs of this chapter, we have tried to outline the advantages, the constraints and the technical possibilities of the use of numerical control.

Exact limits outside or inside which it is economical to abandon the use of numerical control in favour of other solutions cannot be determined. Many of the advantages enumerated in Section 3.1 cannot be costed with precision. The advantage of reducing the time for a repair, of satisfying a request for a replacement part, the avoidance of errors or the rapid modification of a product can hardly be evaluated except in a subjective manner, and not at all by a monetary unit.

On the other hand, when an object is put into production, it is not known how widely it will be used, nor what the chances of making a repeat series are.

Thus it seems that it is not possible to give other than very general indications subject to revision, particularly as the cost of new processes such as numerical control falls rapidly during their initial period of use, and the psychological difficulties of a new technique diminish as a function of the familiarity and confidence of personnel at all levels.

The advantages brought by numerical control are rather different, according to the technique examined, and the limit of profitability can undergo noticeable variations from one system to another. Thus each area must be examined separately. We shall restrict ourselves here to machining proper, leaving aside the other techniques reviewed in Section 3.2.

Drilling and tapping

For drilling, apart from a numerical-control system, there is a choice between three traditional methods: marking out, fixtures or jig boring.

Having eliminated marking out as being insufficiently precise, and the jig boring machine as being almost as expensive and slower than a numerically controlled machine, the choicc lies between a numerically controlled machine and tooling in the form of fixtures carrying bushings to guide the drills and reamers.

If the investments peculiar to one part, represented by a punched tape for a numerically controlled machine and by fixtures for the radial drilling machine, are compared, then it is obvious that the cost of the first is much less than that of the others.

Taking into account the value of machines and their return, the gross cost of executing an operation may be estimated to be a little higher for a numerically controlled machine than for a radial drilling machine with an equivalent capacity, but the expense of preparing a program tape is much less than the cost of the fixtures it replaces.

Under these conditions, it can be shown that the numerically controlled machine is more economic than the tooling and radial drilling machine combination for all runs of between one and a quantity which varies most commonly from 1000 to 2000, but sometimes exceeds 10 000 units.[28–30]

The multihead drilling machine operated by Cardan couplings is a competitor to the radial drilling machine when the quantities to be manufactured reach several thousand units. Thus the range ruled by the radial drilling machine is being reduced at both ends, and perhaps it is not Utopian to think that this machine will lose most of its interest in the fairly near future.

Boring

Boring is performed subject to two constraints of precision, on the diameter and on the position of the centre.

The work on a medium-precision machine requires numerous checks with a plug gauge. Here jig boring machines certainly have advantages, but their capacity scarcely permits the machining of large structures.

As with drilling, tool guides could be used. In this case, the boring machine acts solely as a driving device for the boring bars.

Such tooling is very costly, and its amortizement imposes financial burdens much greater than those which arise from the use of numerical control, at least as long as the run does not exceed several thousand.

Jig boring

In general, the use of a jig boring machine in demanding conditions is avoided, so as not to wear a tool too fast. During the working period of this kind of machine, the cutting time is thus generally much greater than the time for positioning the part relative to the spindle.

For this reason, one could question whether numerical control has an indisputable advantage in this case.

At first, it was thought sufficient to register automatically the manoeuvres carried out by the operator during the machining of the first part in a run. By means of repeating apparatus, a check was made before declamping the completed part to ensure that the operator had not made a mistake in inputting the coordinates. The repeating apparatus then controlled the motions of the machine for the manufacture of the rest of the run.

It was quickly noticed that the use of numerical control in the conventional form, and not as a repeating apparatus, brought, even for a single part, a noticeable economy in time, but that its principal advantage lay in the complete elimination of errors that were ascribable to personnel, and numerical control has been widely applied to jig boring machines.

Turning

Parts of revolution, as they are traditionally conceived, are rarely made entirely from a series of stepped cylinders. Thus numerical-control systems for turning are almost all of the coordinated-movement type.

There is no question, for long runs, of numerically controlled lathes competing with automatic production lathes fitted with form tools or copy lathes with hydraulic followers.

It is only in medium runs and one-off jobs that numerical control finds applications. Otherwise it is competing with copying systems which benefit now from more than thirty years' experience and whose price is modest.

The use of a copier brings constraints unknown in numerical control. The template must first be made and positioned precisely on the template holder. Moreover, if a great quantity of material is to be removed, several passes must be made, which necessitates the construction and setting of an equivalent number of templates. Some shapes do not lend themselves to direct copying; copying machines have a secondary carriage fitted with facing tools, for which further setting is required.

The copying system, although necessitating a rather lower investment than a numerically controlled system, is thus burdened with initial expenses which are renewed at the setting up of each run and which exceed the expenses of making a punched tape.

It seems that the economic upper limit for turning lies between fifty and one hundred parts; beyond this, the hydraulic copy lathe has the advantage.[31–35]

Grinding

In the current state of technology, it is doubtful whether a numerical system would be economical for controlling a cylindrical grinder, as dial gauges are very modestly priced.

The speed of manufacture of prototype camshafts is an indisputable advantage to the credit of numerical control, but it is impossible to reckon this with any exactitude.

If the same method is used for manufacturing limited runs, the construction, setting up and stock-keeping expenses of copiers are avoided. On the other hand, the rapidity of execution is not a noticeable advantage, as, in principle, the work schedule has been established in advance.

As numerically controlled camshaft-grinding machines are considerably slower than mass-production grinders, it has been found profitable to machine masters on a numerically controlled machine, and use them on conventional production machines.

For that reason, direct numerically controlled camshaft grinding is limited to very small series, sometimes less than ten parts.

For the complex parts obtained by three-dimensional grinding, it is much less easy to discern the extent of the economic area of numerical control.

This allows the manufacture of precision parts; failing its use, rough surfaces, forged or precision cast, are used; copy milling is also used. Whatever the method chosen, the parts are finished and polished by hand, a long process which does not ensure great precision.

The output of a compressor or a turbine depends greatly on the exactness of the blading surfaces, so much so that a noticeable increase in the price of blades can largely be compensated for by a slight improvement in performance.

It is on this basis that the usefulness of numerical control may be judged, and not simply on a comparison of the cost price of one type of part.

Drafting and design

As with the preceding process, it is difficult to make a cost comparison between traditional and numerically controlled drafting.

The former is rather slow; much care is needed in the positioning of the French curves, templates and flexible splines that the draftsman uses, and descriptive-geometry diagrams demand a great deal of work.

In other respects, the precision is not outstanding, because the exact form of a spline is not totally defined by the points through which it is to pass. Diagrams often have several hundred points, and some are defined by the intersection of lines making a very small angle between themselves, so that their real position is ill-defined. Unless all drawings are made on painted sheets of metal, they are liable to distortions of the base—paper or plastics film—on which the drawings are kept.

Numerical control has greatly improved the speed and precision of this work. Its main interest lies not in the reduction of the cost of performing these operations, but in the almost total elimination of finishing operations necessary with a manual procedure. These operations call for many subjective decisions, in particular in interpolation and interpretation of data.

The aerospace and shipbuilding industries will undoubtedly remain, for some time, the main application areas of numerically controlled drafting and design.

The principal reason that the automobile industry has become interested in the applications of numerical control to drafting and design work is that it hopes to find in them the means of reducing the lead time in the building of a vehicle, in which the drafting of drawings, the preparation of templates and the construction of masters play a major part. The advantage would not lie in a reduction, perhaps hypothetical, in the cost of these operations, but in the possibility of changing the model more rapidly or making a last-minute modification justified commercially by fashion.

Milling

Numerical control has brought a new solution to the milling of complex shapes, which, until its appearance, were made exclusively by copying. In spite of the excellent development of hydraulic and electrical followers, copying is always marred by errors, and is constrained to use an exact model which is resistant to the contact force of the follower and constructed from a stable material.

Numerical control has caused these inconveniences to disappear, because the error signal required for the operation of the control system is much less than that for a physical follower. There again, the physical model is replaced by a mathematical or numerical representation of the surface to be made. Incidentally, this replacement is sometimes more difficult to achieve than it would appear at first sight.[36,37]

Numerical control is particularly valuable when it is a matter of defining and manufacturing precision mechanical parts such as three-dimensional cams, for which the data are extremely numerous and precise.

The shape of a press tool does not need to be very precise, but the surface of the die must be parallel to that of the punch, and their separation must be equal to the thickness of the blank sheet. As a result, the improvement in precision brought about by numerical control is shown by the reduction in the time necessary for the final adjustment of the tools, some of which weigh more than 20 ton and have a working area greater than 2 m^2 (Figure 3.13).

In the aerospace industry, many composite parts assembled by riveting have been replaced by parts integrally milled from billets of light alloy. The result is a saving in weight and greater reliability (Figure 3.14).

These parts are very highly finished, as any scratch on their surfaces could start crack propagation. They require a great deal of work, because some parts, when finished, are not more than a twentieth of the weight of the rolled or forged billets from which they have been made.

Fig. 3.13. Press tool

In principle, they could be machined by copying a physical master, but, as several passes are necessary, the same number of masters would be needed, which constitutes an almost insoluble difficulty. Thus numerical control today is the universally used method for milling the structural components of aircraft.

Parts such as core boxes, moulds or dies, which are made in more or less large quantities, and for which the required precision is not generally very

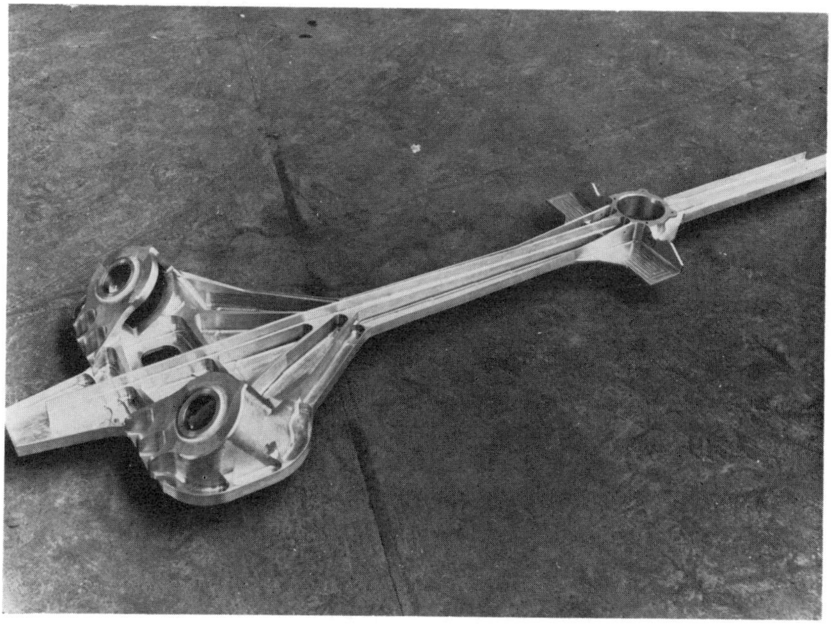

Fig. 3.14. Part of the *Concorde* aircraft

high, remain to be considered (Figure 3.15). The classical copying techniques seem economical since the price of each pattern is amortized over a fairly large number of parts, but numerical control can be of interest again in the manufacture of the pattern, or in the electrodes used in the spark-erosion or electrolytic processes that are beginning to compete with milling.

In summary, the numerical control of two- or three-dimensional milling is certainly economical for the manufacture of the very complex parts found in aircraft structures, in press tools (the adjustment of which needs great care) and one-off parts or parts produced in very small runs.

As for medium-precision parts, and parts made in series, if copying competes with numerical control, then numerical control still has a role to play in the construction of patterns or the electrodes necessary for their manufacture.

Fig. 3.15. Forging die

References

1. W. M. Stocker, 'Proving the profit in numerical control', *Metalworking Production*, 20 December 1961.
2. L. S. Linteroth Jr., 'Economic justification for numerical control', *Automation*, November 1962.
3. 'N.C.: better cost trap', *Production*, December 1963.
4. 'The economics of numerical control', *Machinery*, 26 February 1964.
5. W. P. Krauss Jr. and C. T. Edgar, 'Figuring where NC pays off', *American Machinist*, 8 June 1964.
6. 'Cost reduction through numerically controlled machines (Kostensenkung durch numerisch gesteuerte Maschinen)', *Werkstatt und Betrieb*, **9** (1963).
7. 'Numerical control for economic production', PERA Symposium, 22 and 23 February 1966. PERA Document, Melton Mowbray, England.
8. W. Long, 'Dunlop prove the economic feasibility of tape controlled machine tools', *British Machine Tool Engineering* (publication of the British Association of Machine Tool Manufacturers, London).
9. F. Bell, 'Justifying numerical control in motor vehicles', *Metalworking Production*, 17 May 1967.
10. R. H. Dipple, 'NC. Have you checked these benefits?', *Metalworking Production*, 16 January 1963.
11. 'Punch, notch, nibble with NC', *Tooling and Production*, July 1964.
12. 'Precision tube bending with numerical control', *Tooling and Production*, July 1963.
13. 'Numerically controlled hydraulic bending press, and automatic preselection of the bending angles (Presse plieuse hydraulique à commande numérique et présélection automatique des angles de pliage)', *Ingénieur et Techniciens*, October 1966.
14. 'Wiring terminal panels under numerical control', *Machinery*, 15 August 1962.
15. 'Programmed belt speeds assembly of circuit cards', *Automation*, October 1962.
16. A. H. Johnson, 'Programmed component insertion machines assemble computer logic cards', *Control Engineering*, January 1962.
17. W. A. Hawkins, 'Numerical control balances crankshafts', *Metalworking Production*, 1 August 1962.
18. J. C. Hollande, 'Practical aspects of numerical control applied to turning (Aspects pratiques de la commande numérique appliquée au tournage)', *Méchanique et Électricité*, September 1966.
19. F. J. Robinson, 'Operation of the British Oxygen/Ferranti automatic gas cutting machine', *Machinery*, 3 July 1963.
20. 'Numerically controlled arc-welding machine (Machine à souder à l'arc à commande numérique)', *La pratique des industries mécaniques*, June 1963.
21. 'Numerical control of grinding operations', *Machinery and Production Engineering*, 1 June 1966.
22. Yasumasa Ishihara, 'Camshaft grinding by numerical control', *Metalworking Production*, 2 November 1966.
23. E. Caja and R. Garetti, 'NC grinder makes press-tools', *Metalworking Production*, 5 May 1965.
24. 'Large Cramic tape controlled drafting machine', *Machinery and Production Engineering*, 15 September 1965.
25. D. K. Merris, 'New approaches to computer-aided design', *Product Engineering*, 3 February 1964.
26. R. Gorcey and J. D. Leigh, 'Numerical control improves glass filament winding', *Machinery*, July 1963.

27. 'NC diesinking is here. Are you ready?', *Steel*, 9 November 1964.
28. Profitability of numerically controlled machine tools (Rentabilité des machines-outils à commande numérique)', *Document of the Centre Nationale de l'Automatisation*, 27 May 1963.
29. H. W. Littlewood, 'Practical cost comparisons between numerically controlled and conventional jig drilling', *Machinery and Production Engineering*, 23 March 1966.
30. 'Economic aspects of numerically controlled positional machines', *Machinery*, 5 May 1965.
31. 'Numerical control for turning operations', *Machinery*, 21 October 1964.
32. J. W. Bergman and K. C. Griffith, 'NC lathes turn a quick profit', *American Machinist*, 12 April 1965.
33. 'Lathes head new trends at Brussels', *Metalworking Production*, 8 September 1965.
34. 'Numerical control for center lathes', *Machinery and Production Engineering*, 15 September 1965.
35. 'What case for NC turning?', *Metalworking Production*, 5 October 1966.
36. 'Electronic control of machine tools for contour milling', *Board of Trade Journal* (supplement, 1963).
37. 'Profitability of milling under numerical control at Lockheed-Georgia, (Rentabilité du fraisage sous contrôle numérique chez Lockheed-Georgia)', *Machine Moderne*, November 1964.

Chapter 4

Architecture of the Principal Types of Machine Tools

In Chapter 3, so as not to omit significant developments, our examination of machines included some that are not used for machining proper, and thus lie a little outside the scope of this book. We shall now examine machine tools alone, especially their technical characteristics.

Machines such as boring millers and lathes have not needed any important modifications to be adapted to the requirements of numerical control. Others, such as press drills or radial drills and milling machines are subject to more or less radical modifications. Finally, new machines, which would otherwise never have seen the light of day, have appeared that have been designed specially for numerically controlled operation. Numbered among these are drafting and designing machines, and the multipurpose machines which are described by the term 'machining centre'.

4.1 DRILLING MACHINES

Drilling machines take very different forms, depending on the capacity and precision required of them.

Column drilling machines

These machines are similar to some medium-sized drilling machines (Figures 4.1 and 4.2). Their table is mounted on cross-slides, and the drilling head slides vertically on the column.

The horizontal travel of these machines rarely exceeds 1 m. Some much smaller machines are intended for drilling the printed-circuit boards used in the electronics industry. Most are fitted with four spindles and machine four identical parts simultaneously.

The shape of column drilling machines lends itself rather badly to high precision work, unless the head is clamped to the slide during machining, the feed being obtained by the sliding of a quill. However, the drilling head

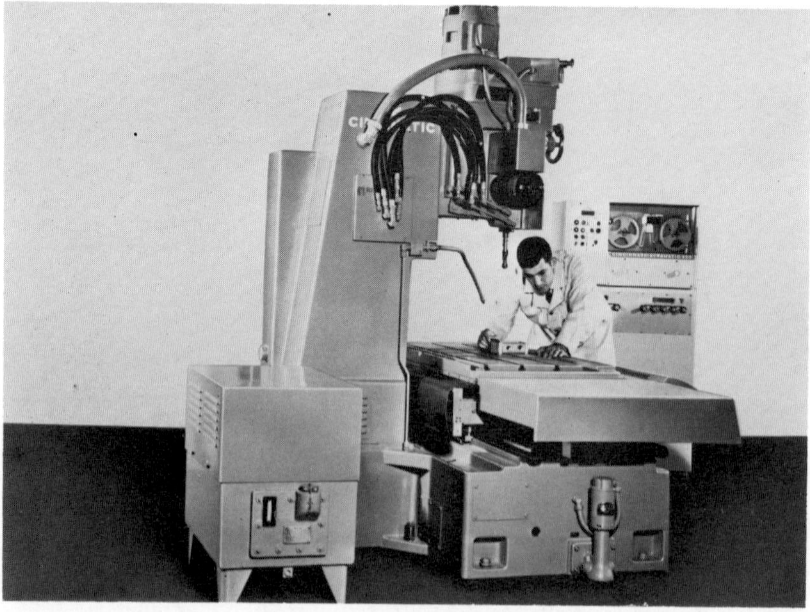

Fig. 4.1. Cincinnati drilling machine

generally includes the spindle gearbox, which causes a rise in temperature during working, giving rise to an expansion which changes the distance between the spindle and the slide. Thus, for precision machines of this type, the spindle and feed gearboxes are generally placed at the top of the column. The torque is transmitted to the spindle by a keyed shaft, and the energy dissipation in the area surrounding the drilling head is reduced to a minimum.

To increase the precision, Hille has placed the axis of the spindle in the plane of the slides (Figures 4.3 and 4.4), thus eliminating the effects of expansion, and reducing the bending couple acting on the head due to its weight and the machining operations. Thus high precision is achieved without resorting to fitting the machine with a quill.

Many manufacturers have built this type of machine, including Cincinnati, Knight, Natco, Pratt and Whitney, Edlund, Olivetti, Jameson, Kolb, Vero, Heller, Huller, G.S.P., etc.

Modification of radial drilling machines

In about 1950, Oerlikon modified some of their radial drilling machines to alter them to precision drilling/boring machines working in orthogonal coordinates. The rotation of the arm is prevented by an auxiliary column,

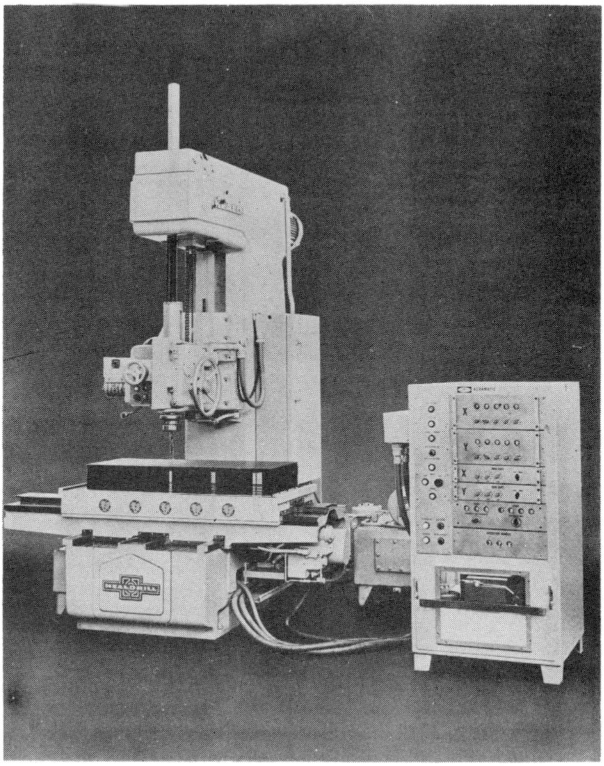

Fig. 4.2. Heald drilling machine

and the part to be machined is carried on a table whose slides are perpendicular to the slides on the arm. The arm is capable of moving up and down on the main column and the auxiliary column, but the machining is performed exclusively by using a quill when the ascending movement of the arm has been locked.

Other manufacturers of radial drilling machines (G.S.P., Jameson, Richards, etc.) have adopted the same principle in building numerically controlled machines (Figure 4.5). In some instances this is only applied to the positioning, and the initiation of the quill movement is controlled manually. The direction of its movement is reversed when a stop reaches a disengaging lever. This system is convenient when a large number of holes of the same depth are to be drilled. There are also machines which have a turret fitted with between six and twelve stops. The travel corresponding to a drilling operation is defined by the registration of a stop, i.e. by the angular position of the turret carrying the stops. It is obvious that, as these positions are limited in number, the method is difficult to use for machining parts which require

Fig. 4.3. Hille drilling/boring machine

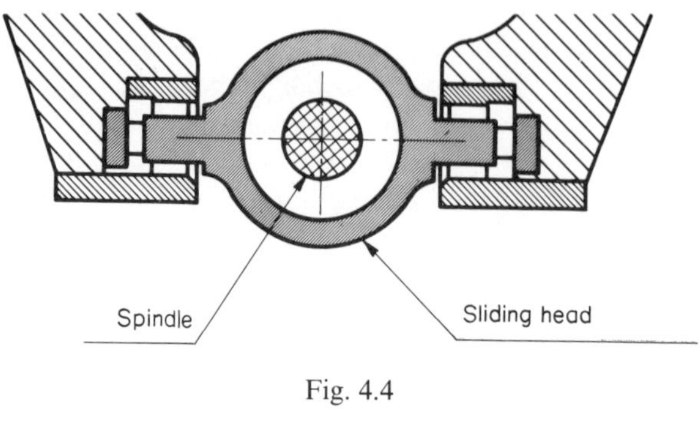

Fig. 4.4

Fig. 4.5. G.S.P. drilling machine

the execution of numerous operations at varying depths and which require the use of many different tools.

Few examples of this type of machine have the vertical position of the tool defined in a truly numerical way.

An 'American' radial drilling machine whose frame was mounted on slides was shown at the 1960 Chicago exhibition; the machinable surface was about 7 m × 1·50 m. This type of machine does not seem to be in widespread use.

Transverse-beam drilling machines

Column drilling machines would quickly reach unacceptable dimensions if one tried to increase their capacity, and larger machines have a form similar to transverse jig boring machines or slabbing millers.[1]

As with radial drilling machines, the ascending movement of the beam is locked during the machining operations. Feeding of the tool is performed by the spindle quill, which slides in the head.

The only difference between these machines and jig boring machines lies in their degree of precision, which affects the dimensions and the design of almost all their components.

Drilling machines of this type have been built by Carlton, Hillyer, G.S.P., Wadkin, B.S.A., Kolb, Jameson, etc. (Figure 4.6).

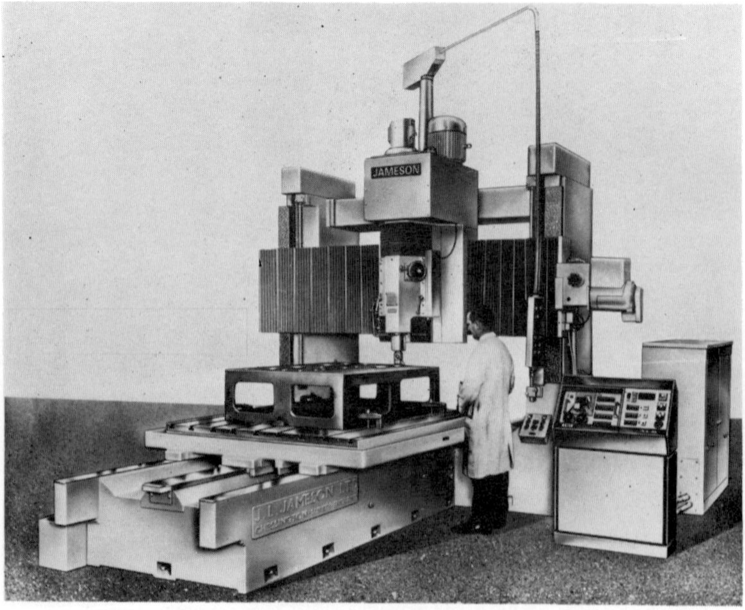

Fig. 4.6. Jameson drilling machine

Moving-portal drilling machines

Large plates such as boiler plates and condenser plates pose a particular problem. Their size necessitates the use of a beam to guide the spindle head, but to avoid having a frame whose length is double that of the part, the only solution is to give the beam a translation movement, with the part clamped on the fixed frame of the machine.

Generally the parts machined on this type of machine are not very thick, even if they are mounted in batches for simultaneous machining. As the beam is part of the portal, and its height with respect to the table is fixed, all the travel of the tool is obtained by sliding the quill in the spindle head.

Drilling machines of this type are, for the most part, designed for machining medium-precision parts. Thus they are generally unsophisticated.

Manufacturers who have entered this field are rather limited in number and include Wadkin (Figure 4.7) and Richards and Jameson in Great Britain.

Fig. 4.7. Wadkin drilling machine

4.2 BORING MACHINES

All types of boring machine are suitable for adaptation to numerical control, whether they have fixed or moving columns, or rotating or fixed tables. However, they have no guidance rests, since tool bars are ill suited to automatic handling and do not fit in to the numerical-control system. The mode of operation is to machine a part successively on its different faces with tools overhanging the spindle. This assumes, in particular, that the rotation of the table is sufficiently precise to guarantee the necessary coaxiality.

In this context, we can cite almost all the names of manufacturers of boring machines: Giddings and Lewis, Lucas, Bullard, Gilbert, Kearns, Cornac, Pégard, Gilly, Scharmann, Froriep, Schiess, Collet and Engelhardt, Wotan, Cerruti, San Giorgio, C.N.M.P., de Vlieg, etc.

4.3 LATHES

Numerical control is applied to parallel lathes,[2–6] and vertical and chucking lathes,[7–12] with the exclusion of single-spindle and multispindle production lathes.

Their general form, as we have seen previously, must fulfil two main conditions:

(a) Successive, and sometimes simultaneous, use of several different tools.
(b) Easy removal of swarf.

The tool changers used can be classified into two different categories: pivoting turrets and removable blocks. Both must fulfil stringent precision conditions. The tolerance in the positioning of a tool is less than half of that permissible on the corresponding diameter.

Pivoting turrets (Heyligenstaedt, Ernault-Somua, Heinemann, Sundstrand, Churchill-Redman, Pittler, Berthiez, Monfors, etc.) are found on the three types of lathe; the difficulty lies in the risk of interference between the turret and the part. The manufacturers have been led to various solutions, depending on whether their machines are designed for machining long, slender parts between centres or short flat parts held in a chuck. The number of positions of the turrets varies from three to twelve.

Instead of using turrets, some manufacturers (Max Muller, Herbert, etc.) have preferred to mount the tools in removable blocks which replace each other during the successive operations required for the complete machining of a part. The technique is similar to the 'Good Chap' turrets currently used in short-run shops.[13]

These units must be placed on the carriage with great precision. Their beds are either V and flat or double V; the tool is set outside the machine, on a station fitted with optical checking devices or dial gauges.

This system has a double advantage; it allows more tools than a turret, and, being set outside the working period proper, it leads to an improvement in the output of the lathe. However, the replacement of one unit by another takes longer than the time necessary to rotate a turret, which, if the tools are numerous, is an inconvenience and difficult to avoid.

The problem of swarf removal has hardly led to noticeable modification of the appearance of vertical lathes or of chuck lathes, which are generally gap bed in form so that the area underneath the part is quite free, and situated on the other side of the slide area. A similar arrangement is not suitable for ordinary parallel lathes, and the most common form is that often found on copy lathes: the slides are in a vertical plane situated behind the line of centres.

4.4 GRINDERS

Centre grinders

These machines are not markedly different from the conventional machines (Landis, Fortuna, etc.).

Camshaft grinders

On conventional grinding machines used in long-run shops, the relative movement of the part and the grinding wheel is the result of a combination of the oscillation of the part holder and the plunging of the spindle carriage. The first motion follows the shape of the camshaft; the second movement controls the conventional phases of rapid approach, roughing out, finishing, sparking out and return.

Numerically controlled grinding machines (Hitachi, Toyo Kogyo) are not based on the same principle. The axis of the part is fixed in space, and the displacement of the grinder carriage causes the movement resulting from the sum of the two displacements. Without entering into too much detail on the control system, it may be said to be driven by pulses sent to stepping motors. There are two tape loops. One relates to the shape of the cams, and its unwinding controls simultaneously the rotation of the part's drive and the plunging of the grinder carriage. One rotation of the part corresponds to one pass through the tape loop. The other loop controls the radial feed cycle, and its commands are combined with those governing the translation of the carriage which results from the shape of the camshaft. The point of contact (Figure 4.8) of the part and the grinding wheel is generally not on the line of centres, and the radius of the grinding wheel is introduced into the equation defining the shape of the cam (see p. 141), which necessitates the most rigorous use of a constant-radius grinding wheel, or at the very least only allows minimum wear of the wheel.

To overcome this inconvenience, Toyo Kogyo have devised a grinder whose grinding wheel undergoes horizontal and vertical displacements such that the tangent at the point of contact (Figure 4.9) is always vertical, with the result that the horizontal travel (Δx) and the vertical travel (Δy) are independent of the radius.

The mechanics of this solution are a little more complex than the other system, since it employs a second carriage with its own control system. It may well be asked, therefore, whether it would not be simpler and more economical to use one carriage with horizontal displacement and to prepare several control tapes corresponding to different grinding-wheel diameters.

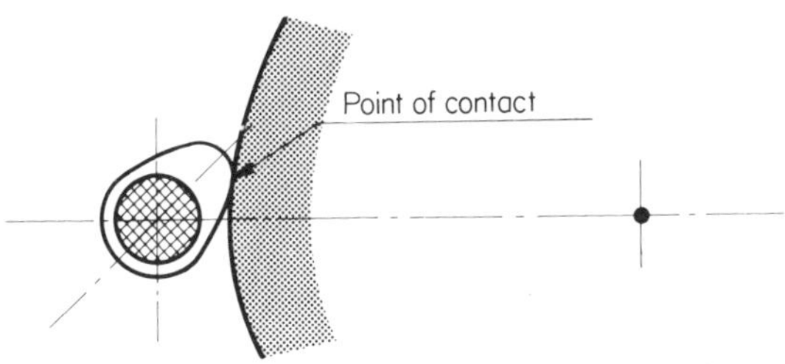

Fig. 4.8

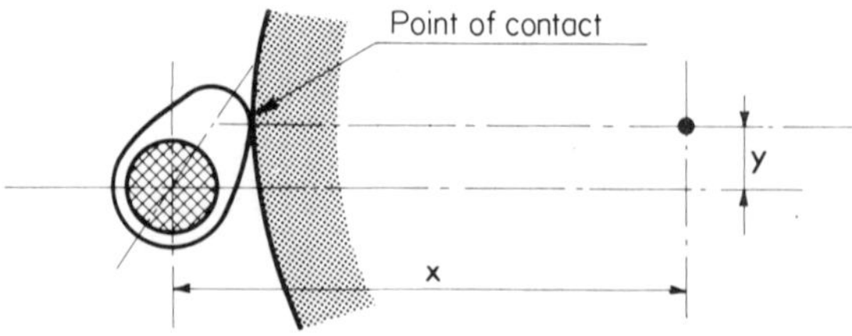

Fig. 4.9

Other grinding machines

Apart from the preceding cases, there are few examples of numerically controlled grinding machines.

Ex-Cell-O have built a numerically controlled milling machine for machining gas-turbine blades. To adapt it to making parts in refractory metal, it was equipped with a spindle carrying a sphero-conical grinding wheel.

The Thomson Grinder Co. have equipped one of their moving-table grinding machines with numerical control, but this attempt does not seem to have been followed up commercially.

4.5 PLOTTING AND DRAFTING MACHINES

The dimensions of most of the modern drafting machines vary between 1 m × 1·50 m and 1·50 m × 7·50 m. The large models are found in the automobile industry, and there are still larger ones in the aerospace industry.

Some [Aristo (Figure 4.10), Contraves, Universal Drafting Machine] are derived from the coordinatographs used for drawing geographical maps. They take the form of a horizontal table, with a transverse beam running on prismatic slides. A system of racks and gearing synchronizes the movements at its two ends. On this beam sits a carriage carrying the stylus holder or optical apparatus with a reticule used to centre on points whose coordinates are to be measured.

If the machine exceeds a certain size, it is difficult for an operator to centre on points towards the middle of the working area. The viewer must then be replaced by a television system.

Other machines resemble the large drawing boards used for car-body lines. Their face is inclined by about 10° to the vertical. Some have a light carriage carried on a slide at the top of the machine; others [Cramic, D.E.A., etc. (Figure 4.11)] have an extremely rigid beam at the bottom which guides the carriage. With this type of machine, the use of a television system is less necessary than with the preceding type.

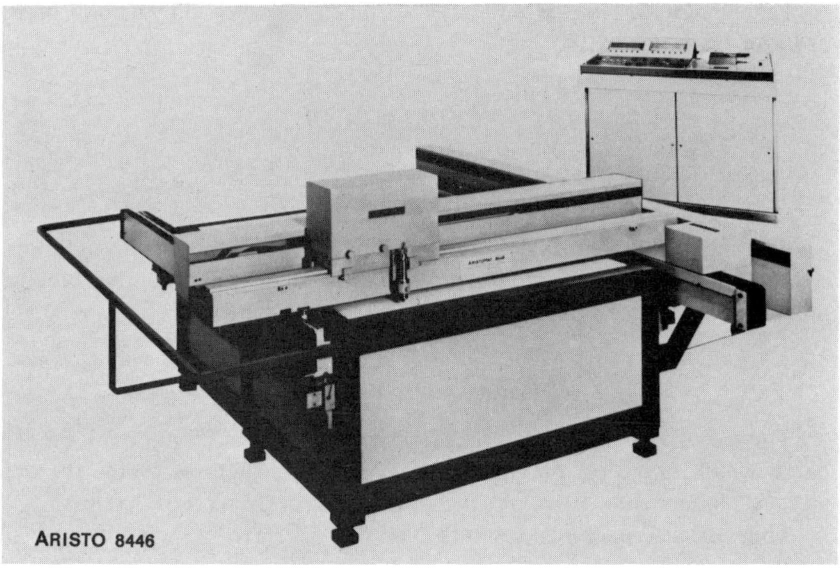

Fig. 4.10. Aristo drafting machine

Fig. 4.11. Kongsberg drafting machine

Finally, Gerber have built tiltable machines with dimensions of the order of 1·5 m × 7·5 m.

4.6 MILLING MACHINES

Conventional milling machines

All types of conventional milling machine have been fitted with numerical control without any fundamental modification, but it seems that the machines most readily adapted have vertical or swivelling spindles.

Portal milling machines

The aerospace industry makes considerable use of integrally milled parts, which make up the spars and panels of structures. Some of these parts are longer than 10 m, but their thickness rarely exceeds 300 mm.

Thus special machines for this type of part have been made, and are called spar-milling or skin-milling machines. Some (Cincinnati, Onsrud, Hillyer, Morey, etc.) have the form of slabbing millers. Sometimes their bridge carries two or three machining heads whose transverse displacements are controlled independently. This allows the simultaneous machining of several identical or symmetrical parts.

Others (Cramic, etc.) have a fixed table and a moving portal, which, for an equal capacity, makes them notably smaller than the preceding machines.

Each end of the portal is fitted with an independent measuring system. Thus differences in displacement are compensated, and the portal is maintained perpendicular to the slides, guaranteeing excellent precision despite the dimensions of the machine.

4.7 MULTIPURPOSE MACHINES

As there are considerable gains in performing the greatest possible number of operations on one machine and in one mounting to reduce lead times in manufacture and to avoid paying for several setting operations, it was quickly realized that conventional machines had to be equipped with systems that were capable of controlling drilling, tapping and contour milling with equal efficiency.

If the machines in question are of the column drilling type or millers with vertical spindles, they do not readily allow successful working, except on flat parts. Thus machines with horizontal spindles fitted with rotating tables (Kearney and Trecker, Cincinnati, G.S.P., etc.) capable of machining holes parallel to the plane of the table were rapidly turned to. The rotation of the table is in steps of 45° or 90°, depending on the intended purposes, unless a rotation defined in thousandths of a degree is preferred[14] (Figure 4.12).

This type of machine will not, however, cut holes that are oblique to the table, and for some parts is an awkward limitation.

For this reason, machines have been made with rotating tables whose spindle is inclinable (Sundstrand OM III) (Figure 4.13) or whose table surface has two orthogonal axes of rotation (Kearney and Trecker, Milwaukee-Matic 3 and 5) (Figure 4.14). These machines can carry out five types of movement, each position of the tool being defined by five coordinates, three of which are linear and two of which are angular. In practice, if the spindle is inclinable, the machine is also often provided with a quill, because it would be difficult to effect the translation of a drill parallel to an oblique axis by combining two orthogonal translations. Machines of this type are sometimes said to have six axes.

For an equal capacity, a five-axis machine has much greater external dimensions than a three-axis machine.

For parts with large dimensions, machines whose moving column rotates and carries a spindle which is itself inclinable have been made (Figure 4.15).

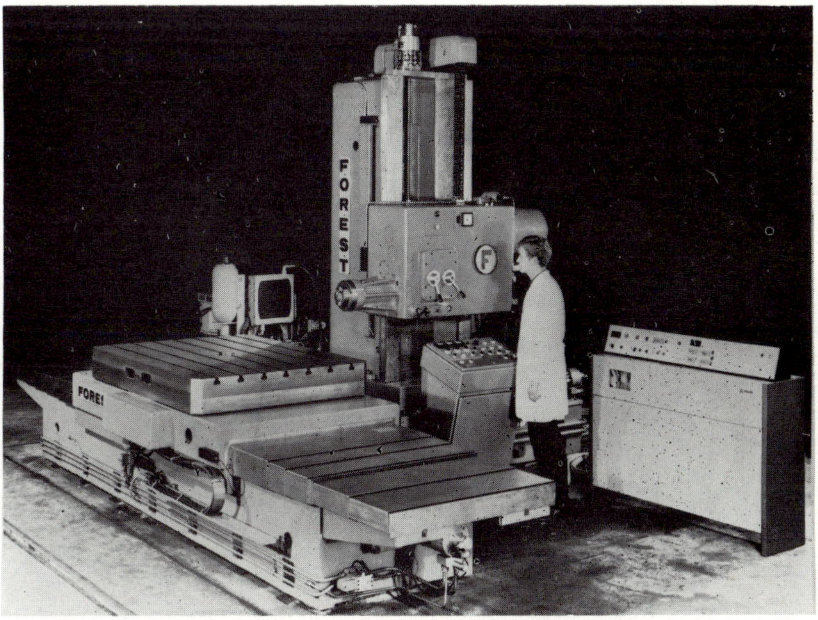

Fig. 4.12. Forest machine tool

Fig. 4.13. Sundstrand OM III machine tool

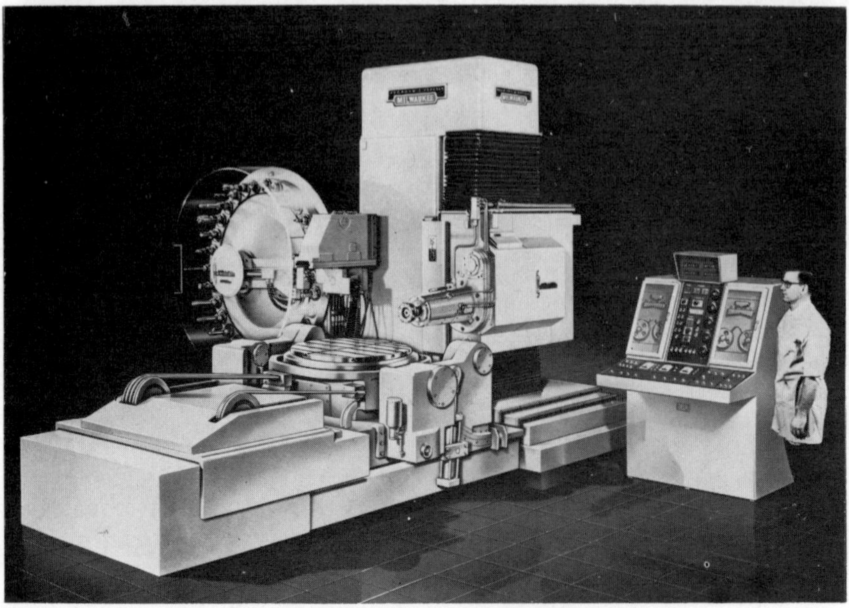

Fig. 4.14. Kearney and Trecker Milwaukee-Matic machine

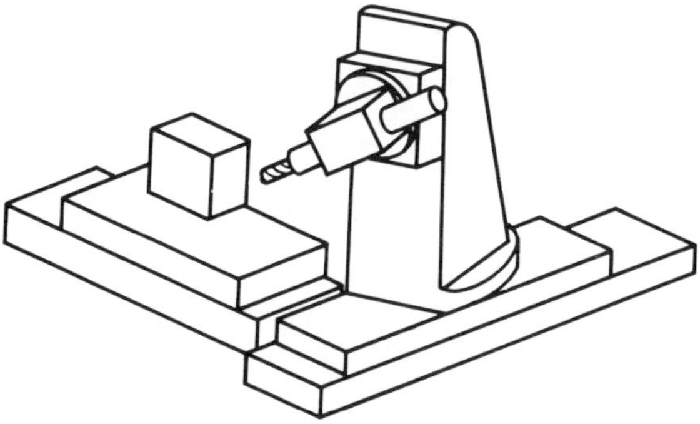

Fig. 4.15

The Hughes machine manufactured under licence by Burkhardt and Weber (Figure 4.16) must be mentioned. The part is fixed to a rotating table which moves in two directions. Three machining units perform, respectively, drilling, milling and precision-boring operations. The first and the last units are fitted with automatic tool changers.

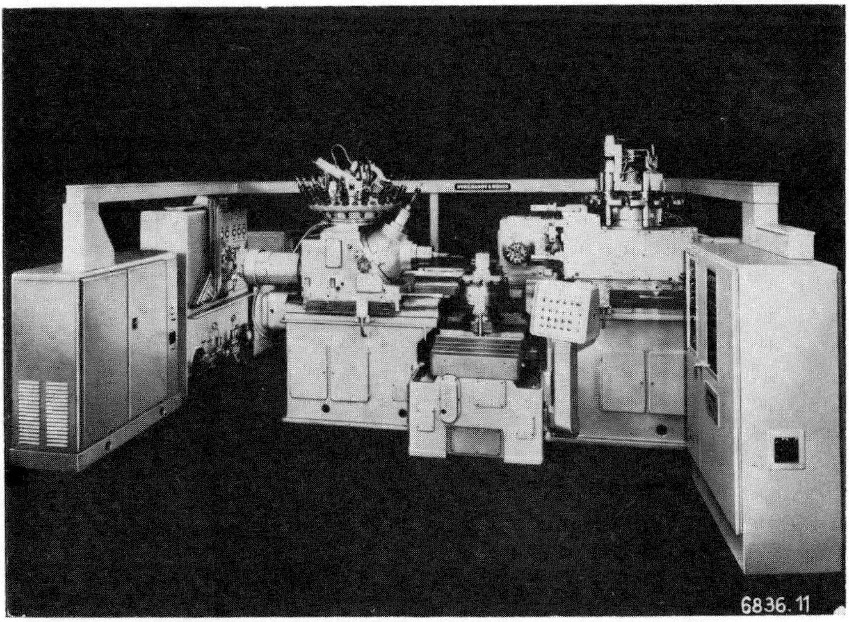

Fig. 4.16. Hughes–Burkhardt and Weber machines

4.8 ACCESSORIES

As the output of numerically controlled machines is of prime importance, many are fitted with accessories that are designed to reduce the dead time.

Turret tool holders

Many drilling machines, and even one boring machine (Wotan), have been fitted with a turret with five, six or eight spindles. Each of these carries a particular tool, and thus a series of drilling, spot-facing, chamfering, tapping, etc., operations can be carried out successively at the same point and without loss of time[15,16] (Figure 4.17).

For each position of the turret, there is a corresponding speed of rotation, an advance, a travel and a cycle.

So that the quality of the machine should not be seriously reduced, the indexing of the turret must be extremely precise. The Gleason *curvic coupling* system is used, for example.

However, when several spindles are arranged in a limited space, it is necessary to shorten them and reduce the distance between their bearings, but this does not improve the precision. This is why some manufacturers, instead of making the axes converge towards the centre of the turret, arrange

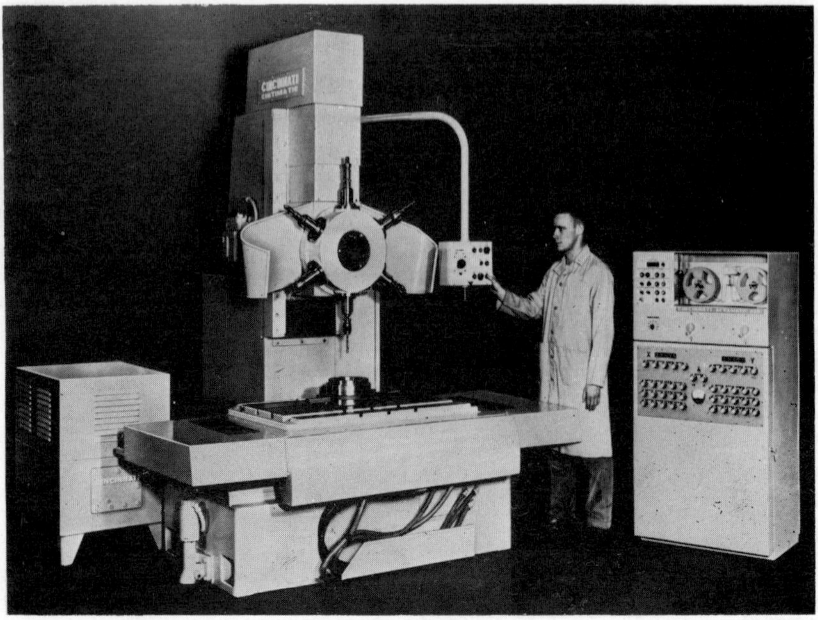

Fig. 4.17. Cincinnati machine

them tangentially to a circle, which allows, for an equal space, a slight increase in the distance between the bearings (Figure 4.18).

This solution is not universally applicable, since, for machining, many parts require more than six or eight tools. Moreover, it is ill suited to machining deep pockets, since the tools carried by spindles adjacent to the spindle in use risk coming into contact with higher areas of the part during machining (Figure 4.19).

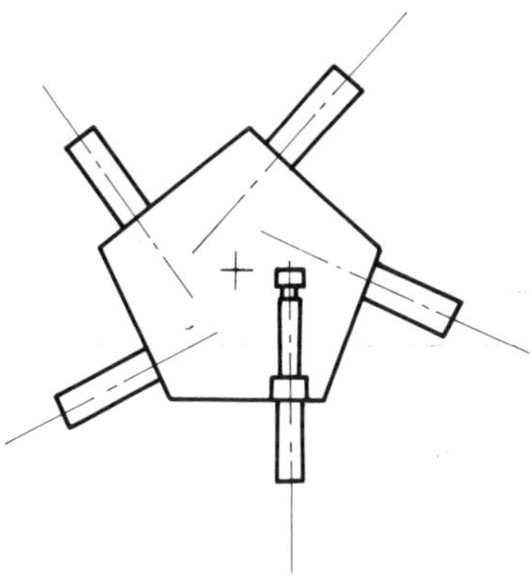

Fig. 4.18

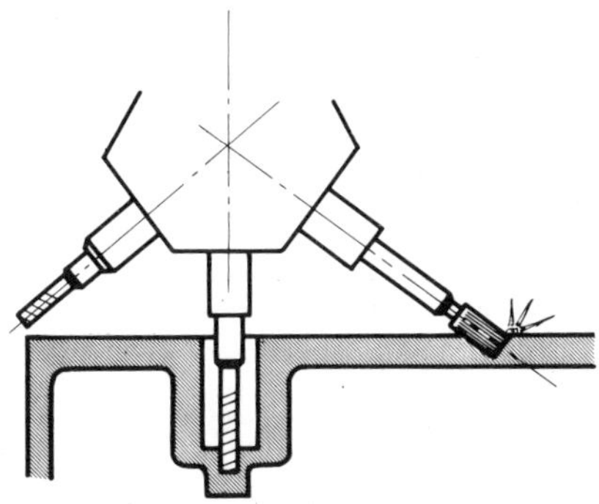

Fig. 4.19

Thus this type of equipment is especially suited to making flat parts which do not need a great variety of tools.

Tool magazines

A turret holds, of necessity, a very limited number of spindles; to remedy this defect, the idea of fitting machines with a magazine containing a much greater number of tools and a changer capable of substituting one for another in the nose of the spindle has been conceived.

The time required for this operation is not always shorter than that required by an operator for the same task, but the apparatus has the advantage of operating without delay, whereas the operator is not always immediately available. Moreover, when the machine is rather large, the spindle would have to be lowered to within reach of his hand. The weight of a large tool, a surface-milling cutter or a boring bar can exceed 40 kg, and mechanical or manual help would be required for its positioning.

The mode of operation of a magazine machine differs from that of a turret machine. First, it executes all the operations requiring one particular tool, then the change is effected, and the machine returns, if necessary, to the same points to perform subsequent operations with another tool.

Magazines have extremely diverse forms and capacities. The problems to be resolved are varied, but the manufacturers have tried not to encroach upon the patents of their competitors.[17–20] Some of the most representative of the various solutions will now be briefly described.

One of the most ingenious is the Olivetti Auctor. A drum containing a dozen different tools positions them one after another before the spindle. This deposits the tool in one of the cells of the drum and, after rotation of the magazine, grasps and fixes the next tool (Figure 4.20).

Another type, which can hold twenty or so tools (Kearney and Trecker, Cincinnati, G.S.P., etc.), is fixed on the spindle head and follows it during displacements (Figure 4.21). At the moment of exchange, a double arm simultaneously grasps the tool in the spindle and the tool in the magazine, which is brought opposite the arm by pivoting the housing. A rotation of the arm performs the tool change.

Magazines with a much greater capacity would scarcely be able to follow the spindle head during its movements. They are sited near to the machine (Sundstrand OM III, Figure 4.13), and there is a carriage which, after having grasped the tool, brings it sufficiently close to the spindle to perform the substitution. The spindle must, in general, be moved to a well-defined position, which lengthens the dead time a little.

Fig. 4.20. Olivetti Auctor machine

Fig. 4.21. Kearney and Trecker magazine

Giddings and Lewis (Figure 4.22) have arranged two rotating magazines on the spindle head of their vertical-spindle machine. Each of them has an arm which carries the tool from its cell to the nose of the spindle or vice versa.[21] As the odd-numbered tools are in one magazine and the even-numbered tools in the other, the substitution is very rapid, and the magazine containing the tool waiting to be substituted is rotated while the previous tool is machining.

Avey, Herbert, etc., (Figure 4.23), use a head carrying two opposite spindles. While one is working, the other, which does not turn, receives the following tool. The substitution of one tool by another only requires the time needed to complete a half revolution of the head, which is very rapid.

The capacity of a magazine does not necessarily limit the number of tools intended for the manufacture of a part. The operator can, during work, remove from a magazine those tools which are no longer needed and replace them by others.

The instructions given to the machine include the identity of the tool to be used. This is indicated, in some cases, by rings of different diameters which form a binary or binary-decimal code on the body of the tool holder (Figure 4.24). Electrical switches read the information when the magazine drum is set in rotation, and stop it in the position corresponding to the tool to be exchanged.

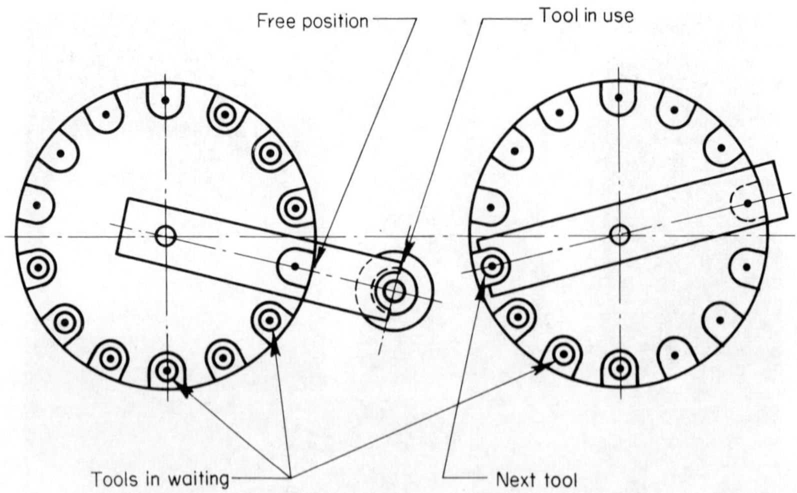

Fig. 4.22

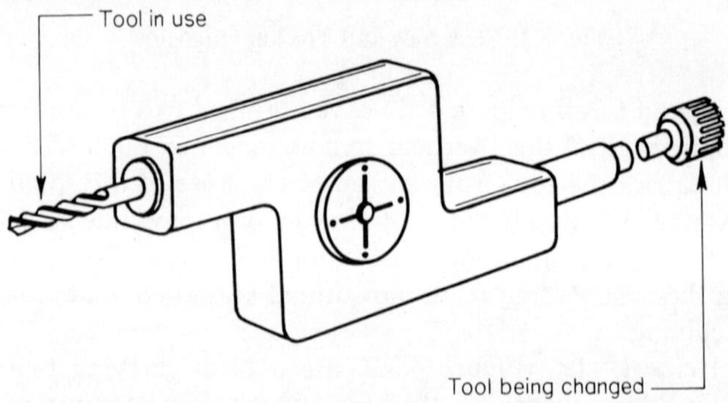

Fig. 4.23

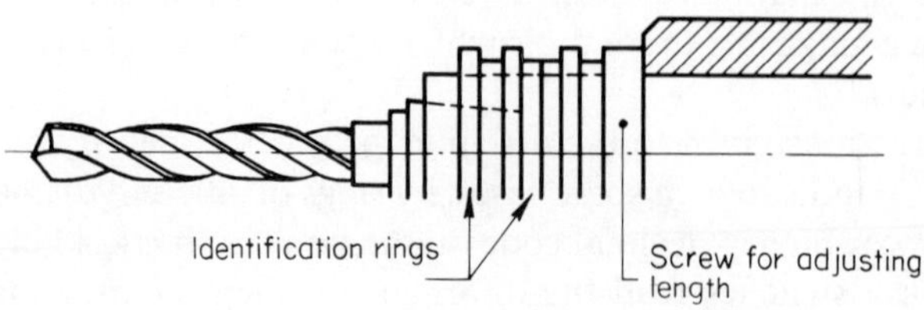

Fig. 4.24

Other systems of identification make use of a coded key supplied at the same time as each tool. Next to each housing intended for a tool, the magazine carries a slot in which is inserted the key which activates the reader.

This method reduces the overhang of the tool spindle, but it necessitates the return of the tool to a predetermined housing, which, in comparison with the previous system, takes a little longer and requires more complex control equipment. However, if the system is well planned, all the operations of selecting and restoring the tools are carried out during the periods of machining, and only the replacement of one tool by another constitutes real dead time.

Lastly, there are systems in which the tool is only identified by the number of the cell in which it is placed. Obviously the operator must not make any mistake when he places the tools in their housing, as each mistake could have unfortunate consequences.

Tool holders

To obtain high precision, it is vital that the axis of a tool, reamer or boring bar should coincide with that of the spindle.

Many machines use large-angle cones of the U.S.A. type, which are not liable to jam under the effects of clamping forces or tool reactions. They are drawn in by means of a central rod operated by a pneumatic or hydraulic jack, unless the grip is applied by a spring or Belleville washers. The slack having been taken up, a premature drop in feed is avoided. With these cones, the conventional method of driving is by two opposed lugs. This necessitates the orienting of the tool to suit the spindle, and, to avoid this, drives have been made in the form of chamfered toothed gears; this allows them to engage in all positions.

One solution, employed in particular by Sundstrand in their OM III and OM IV machines, uses tools with cylindrical shanks driven by adhesion using an elastic cone similar to a Jacobs chuck; this assembly has excellent precision (Figure 4.25).

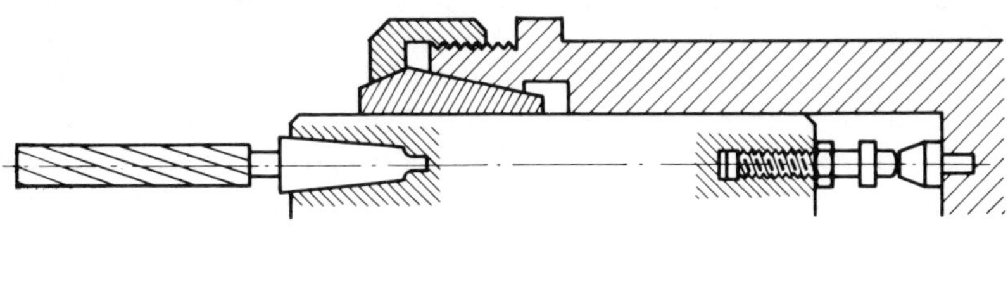

Fig. 4.25

Presetting of tools

The importance of presetting justifies the use of precise measuring apparatus fitted with optical attachments or comparators and standard gauges by which the radial and longitudinal position of a cutter edge with respect to the axis of rotation of the spindle is checked.

The bars and cylindrical tool holders are fitted, at the rear, with a counter-screwed stop which butts against a bush at the bottom of the housing inserted in the spindle (Figure 4.25).

If the spindle is fitted with a cone, a bar holder carries a cylindrical hole into which the bar fits. On the outside diameter, a trapezoidal thread engages a locking screw which acts as a stop which is adjustable for length (Figure 4.26); the longitudinal and radial adjustments are effected in succession.

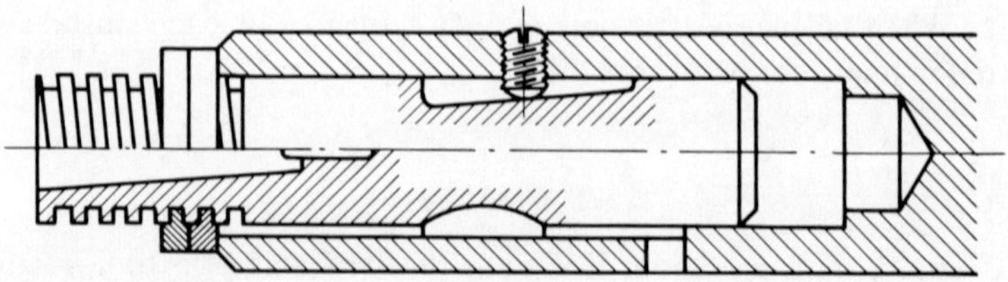

Fig. 4.26

There are also some boring bars fitted with Morse cones which act as drill holders, but this equipment can hardly be used for guiding large-diameter fixed boring bits. In this case, tools with floating cutters which centre themselves automatically in a previously roughed-out hole whose axis is correct but whose diameter is not necessarily accurate are used.

Workpiece changers

To locate and clamp a part suitably takes some time, and a way of performing this task when the machine is still working on the preceding part has been sought. The part is fixed on a standardized pallet, and it is this assembly which is placed on the table of the machine at the last moment, at a place defined in advance by stops against which the pallet is clamped.

On some machines (Cincinnati, Sundstrand, etc.), the workpiece pallet is mounted in readiness on an extension of the bed (Figure 4.27), and there is an automatic system which positions it and clamps it on the table in a few seconds. The new task begins immediately, provided that the corresponding tape is loaded and a start has been made, during the last phase of the preceding job, on replacing the tools which have already been used by those necessary for the new job.

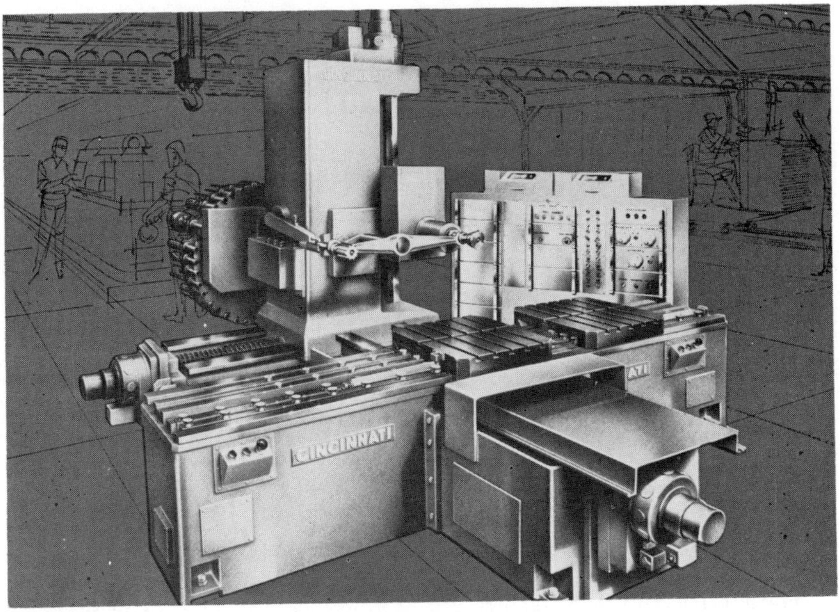

Fig. 4.27. Cincinnati machine

On moving-column boring machines, the movement of the column covers two work stations (Figures 4.28 and 4.29).

The setting up of a large part is a sufficiently long and delicate operation for the improvement in output to justify the additional investment represented by the second table and the lengthening of the slide carrying the column.

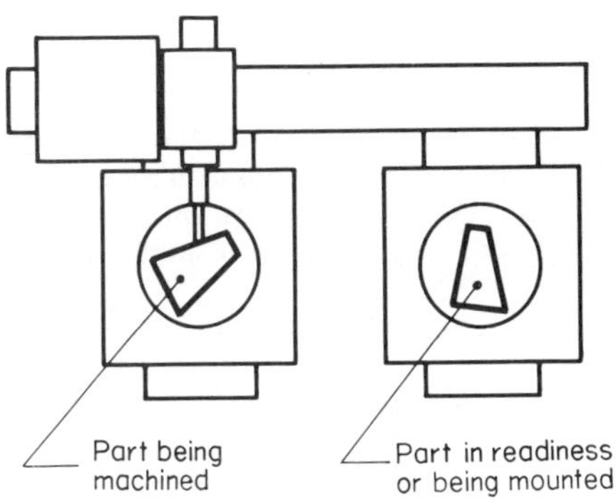

Fig. 4.28

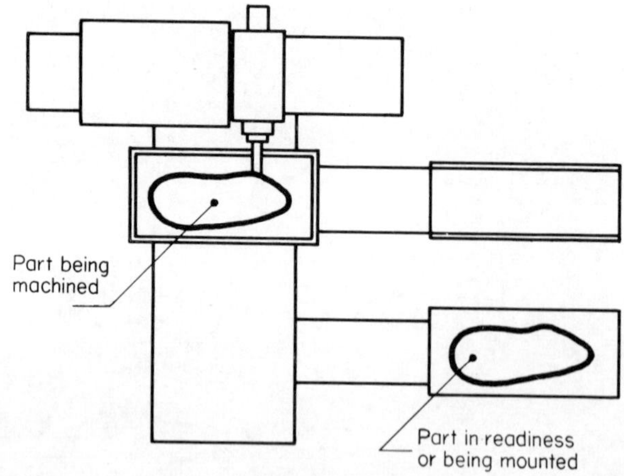

Fig. 4.29

Control consoles

The control console is the means of communication between the operator and his machine. As the cost of machines varies in the ratio 50 : 1, the console, according to the application, can present rather different appearances, ranging from the most basic to the most complete.

For a very simple machine, the console is a box carrying a rather limited number of dials and buttons. On the console of a sophisticated machine, the accessories are much more numerous, and may include tape readers, indicator lights and measuring devices. To these are added the switches necessary for the input of origin shifts, and those by which a coordinate may be added to any of the axes, or the direction of one of them inverted to produce a symmetry.

References

1. 'Jameson special purpose', *British Machine Tool Engineering*, No. 1, 1964.
2. R. Green, 'The "S pilote" numerically controlled lathe (Le tour "S pilote" à commande numérique)', *Machine Moderne*, May 1966.
3. 'Communiqués, exhibitions and lectures (Communiqués, expositions et conférences)', *Microtechnic*, **XX,** No. 3.
4. 'Ikegai A-20 numerically controlled lathe', *Machinery and Production Engineering*, 7 December 1966.
5. 'New British lathe designed completely for NC', *Metalworking Production*, 3 May 1967.
6. 'General purpose NC turning "the Numericon"', *Machine Tool Engineering*, No. 3, 1967.
7. BSA Type 95E chucking automatic with numerical sequence control', *Machinery*, 28 October 1964.

8. 'East West team up on NC lathe', *American Machinist*, 23 May 1966.
9. 'The Csepel numerically controlled 10 station turret lathe', *Machinery and Production Engineering*, 2 July 1966.
10. 'Liechti DR 650 programme controlled chucking lathe', *Machinery and Production Engineering*, 19 October 1966.
11. R. Iredale, 'Which way now for NC turning?', *Metalworking Production*, 8 February 1967.
12. 'Herbert 3 and 4 Programauto N/C lathes', Catalogue No. L 1241, Alfred Herbert, Coventry, England.
13. 'Herbert turning centre', Catalogue No. E 760, Alfred Herbert, Coventry, England.
14. 'Closing the gap', *Steel*, 13 September 1965.
15. 'Twenty tools without a changer', *Metalworking Production*, 15 September 1965.
16. 'Vero Autodrill series 1000 tape controlled drilling machine', *Machinery*, 4 March 1964.
17. 'The automatic tool change', *Steel*, 25 May 1964.
18. 'Automatic tool changing', *Machinery and Production Engineering*, 8 September 1965.
19. 'Chain increases NC tool changer capacity', *Metalworking Production*, 6 April 1966.
20. 'New Milwaukee-Matic has chair-type tool magazine', *Metalworking Production*, 31 May 1967.
21. 'Tool changer magazines move with vertical head', *Metalworking Production*, 1 September 1965.

Chapter 5

Job Preparation

As a job on a numerically controlled machine is, in general, prepared in advance and is liable to be repeated, it is important to be able to choose at the last moment the machine on which it is to be carried out. It is also important that the information media should be transmissible from one factory to another, and even from one company to another. Standardization thus plays an important rôle in this area.

The problems posed to the programmers are very different, according to whether it is a question of working point to point, or contouring, and for this reason these two types of operation will be the subject of separate sections.

The languages available to programmers are numerous.[1–7] Some are very comprehensive, and can solve the most complicated problems; others are simpler and more economical because they have been given more restricted capabilities. They cannot be described here in detail and only their general characteristics will be mentioned.

If a point-to-point program consists of only a few operations, it is possible to perform all the calculations by hand, and to write the instructions, but as soon as the number becomes large, the use of a computer is inevitable. In practice, contouring jobs always make use of a computer.

5.1 STANDARDIZATION

The creation of an automatic programming system is an important undertaking which involves very large sums of money and which requires highly qualified specialists.

Many companies, each working alone, have developed languages adapted to the products they manufacture. The inconveniences of this disparity quickly became apparent, since the users found themselves obliged to use several methods in parallel, complicating the training of their personnel.

Moreover, the results of the calculations were expressed in a form which was not equally readable by any machine.

The data for problems given to machine and numerical-control-system manufacturers are extremely varied, and it is impossible to find a solution which is at the same time universal and economical.

However, great efforts have been made by international organizations such as the I.S.O.* and the C.E.I.† to establish compatibility between all the systems, and a certain degree of interchangeability between the instruction media they use.

Among the recommendations already published are:

ISO R 840	Code for numerical control of machines (compatible with the I.S.O. 7 bit character set).
ISO R 841	Axis and motion nomenclature for numerically controlled machines.
ISO R 1056	Punched-tape block formats for the numerical control of machines, coding of preparatory functions G and miscellaneous functions M.
ISO R 1057	Interchangeable punched-tape variable block format for positioning and straight-cut numerically controlled machines.
ISO R 1058	Punched-tape variable block format for positioning and straight-cut numerically controlled machines.
ISO DR 2539	Punched-tape variable block format for contouring and contouring–positioning numerically controlled machines.

The work in progress concerns the programming languages, the formats for contouring operations, magnetic tapes and graphical symbols.

5.2 POINT-TO-POINT PROGRAMS

To produce a program automatically, a certain amount of basic information must first be gathered and sent to the computer, generally by means of a deck of punched cards.

Geometry of the tool

The shape of the cutting edge of the tool is defined, as well as its position with respect to the spindle. The description is standardized and allows the description of a tool with a relatively complex shape.

Tools so defined are drills, reamers, taps, lathe tools and cylindrical, toroidal, conical or cylindro-conical milling cutters.

* International Standards Organization.
† Comité Électrotechnique International.

Breaking down of operations

When a machining job is given on a drawing by a line such as that in Figure 5.1, the part programmer breaks it down into elementary operations such as, for example:

	ϕ	Depth
1. Centering	8	6
2. Drilling	10·25	42
3. Drilling	12·5	16
4. Tapping	12 × 1·75	30
5. Reaming	15·5	13
6. Spot facing	24	4
7. Reaming	16 H^7	12
8. Chamfering	22, $\alpha = 45°$	5

The dimensions relating to operations 1 and 5 are not apparent on the drawing; it is up to the part programmer to judge whether it is necessary to insert them, and, in this case, to define the tools to be used.

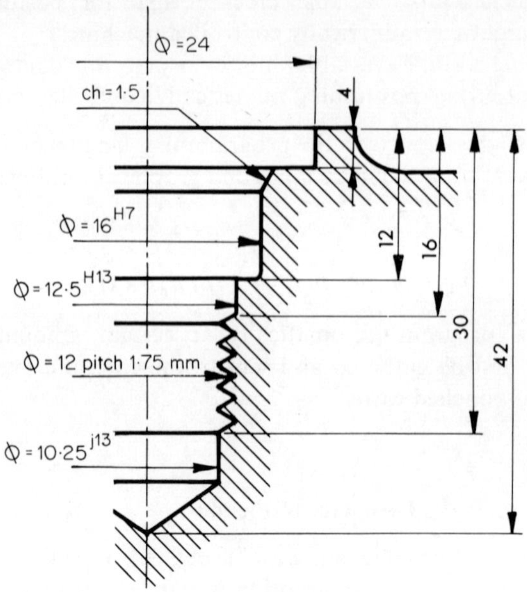

Fig. 5.1

Definitions of the locations of axes

The information regarding the location of an axis is sent to the machine in the form of the orthogonal coordinates of its projection on a plane which is perpendicular to it.[8]

DEFINITION OF A POINT

On the drawings of a part given in a traditional way, not all the axes of the boring operations are directly defined by coordinates, and their projection is sometimes situated at the intersection of two curves which are, save for rare exceptions, straight lines or arcs of circles, since the ruler and compasses are the traditional instruments of mechanical designers.

If the curves have two intersections, the part programmer specifies whether the selected point corresponds to the larger or the smaller value of one of the two coordinates, which is sufficient to avoid ambiguity.

DEFINITION OF A STRAIGHT LINE

To define a straight line and the positive direction associated with it, one can use:

(a) Two points, the positive direction going from the first to the second.
(b) A point and a direction. This is defined as making an angle, given by its algebraic value, with another directed straight line. A coordinate axis is a particular case of a directed straight line.
(c) A point and a circle to which it is tangent. The problem has two solutions. The ambiguity is removed by specifying whether it is the line which makes the maximum or minimum angle with one of the coordinate axes.
(d) Two circles to which it is a tangent. The solution chosen from the four possibilities is indicated, for example (Figure 5.2), by saying that it

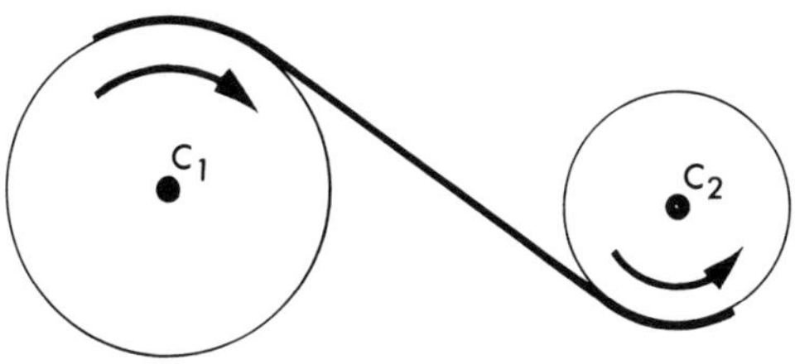

Fig. 5.2

blends with the first (C_1) in the negative direction and the second (C_2) in the positive direction.

DEFINITION OF A CIRCLE

A circle can be defined in many ways, taking into account its geometrical properties. Not all are used in current languages, and only those which are used frequently will be mentioned. The following are various ways of defining a circle:

(a) A circle which passes through three given points.
(b) A circle which has a given centre and radius.

(c) A circle of given radius which is a tangent to two straight lines. Given that the lines are directed, it is specified whether the centre is to the right or left of each.

(d) A circle of given radius which is a tangent to two arcs or to an arc and a straight line (Figure 5.3).

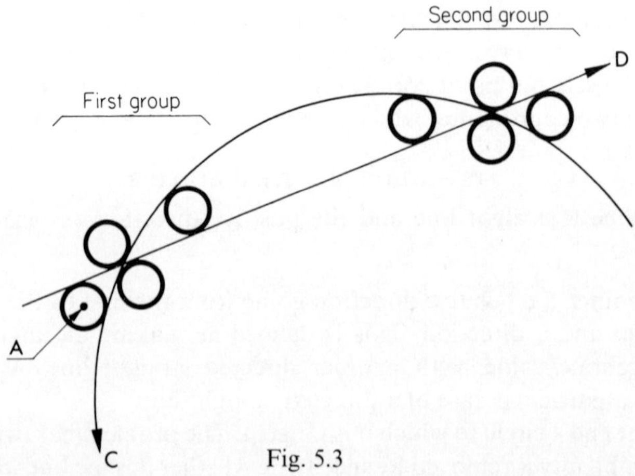

Fig. 5.3

The same information is given as above to indicate whether the circle required belongs to the first or second group of four which is met in following one of the curves in a given direction. If, where two possibilities exist, one of the coordinates of the chosen centre is greater than that of the other, this is an alternative means of identification. For example, the arc A is to the right of D, to the right of C, and in the group lying at the intersection of C and D which has the minimum x value.

(e) A circle which is a tangent to a straight line at a given point and a tangent to another circle. We specify whether the centre is to the right or left of the straight line.

Many other cases could be imagined. They would arise more frequently from mathematics than from mechanics, and the need for their solution occurs rarely, but it is always possible, if not economical, to write a program specially to solve a problem whose solution has not already been programmed.

Technological data

Normally it is the part programmer who determines the cutting conditions—cutting speed, feed, coolant—by means of tables and charts, and by making the necessary assumptions on the machineability of the material to be used, the casting or forging allowance, etc.

In addition, many control systems leave to the operator the possibility of increasing or decreasing the speeds when he notices that the real conditions are different from those which the part programmer had foreseen.

For some time certain programming languages (EXAPT, IFAPT, etc.) have incorporated the possibility of automatic determination of the working conditions as a function of theoretical data known to the operator—the type of material, the geometry and material of the tool, the power of the machine, etc.[8–10]

For several years attempts have been made to improve the output of machine tools, particularly those which are controlled by a numerical system.[11,12,26–37]

The working conditions are rarely constant during an elementary operation. The machineability of the metal, the condition of the edges of the tool, the casting allowance and the size of the cut all vary.

When the cutting speed and feed are invariable during a cut, they must be chosen as a function of the most unfavourable conditions which can be met. If they are constantly adjusted to the real conditions and to the intended goal, there will be a resulting improvement in output. This method is designated by the term 'adaptive control'.

It would be desirable to take into account the quantities which express the power absorbed, the torque and the bending moment applied to the spindle, the temperature of the cutting edges, the vibrations of the machine and of the part, as well as the macro and microgeometric dimensions (roughness) of the surface during machining.

Optimization of different properties such as the speed of metal removal during the roughing stage, dimensional precision, surface finish and tool life (to avoid interruption of work during a long finishing cut) may be proposed.

The variables which may be influenced are cutting speed, feed rate and, sometimes, depth of cut.

In practice, some measurements, such as the temperature of the cutting edge, or the surface finish, are difficult to make, and measurements are often limited to the power, to the torque or the bending moment, the feed rate being measured rather than speed of cut.

However, the possibility of notable improvements in adaptive control, due to progress in transducers and the better understanding of the mathematical description of the cutting process, must not be excluded.

5.3 CONTOURING PROGRAMS

The curves and surfaces produced by means of numerical control may be classified in three categories:

(a) Curves and surfaces made up from the juxtaposition of segments belonging to mathematical curves and surfaces.

(b) Non-mathematical curves and surfaces obtained by fitting them to experimental points, and which must be reproduced faithfully within strict tolerance limits.
(c) Non-mathematical curves and surfaces, the production of which does not require adherence to very strict tolerances.

The majority of shapes of mechanical parts are found in the first category. With very few exceptions, they are designed using straight edges and compasses, and the curves consist of straight lines and circles. Surfaces consist of the plane, the cylinder and the cone of revolution, the sphere, the torus, the quadratic surfaces and some ruled surfaces. If, in addition, a normal, conical or developable helicoid is included, it is used to generate screw threads and the flanks of gear teeth, and numerical control is very seldom used to produce these shapes.

The second category comprises all the surfaces obtained by fitting after experimentation. They are, for example, the shapes of ships' hulls, the fuselages and wings of aircraft, or the blades of turbines. They have been carefully improved to obtain an optimal fit, and the quality of the products obtained depends principally on the fidelity of the reproduction.

These surfaces are purely experimental, and cannot be interpreted *a priori* except by measuring coordinates on a close grid of points.

The last category is made up of the surfaces, such as automobile-body shapes, which are created in a purely intuitive manner to satisfy aesthetic notions. Their reproduction is not liable to such strict tolerances as the surfaces in the preceding category.

The general problem divides into two, according to whether a curve or a surface is to be produced. Curves embrace the operations of drafting, flame cutting, continuous welding, or the production of surfaces defined by one curved line which can be the generatrix of a ruled surface. Surfaces are involved in the machining of the majority of those surfaces which do not belong to the preceding category.

Plane curves

MATHEMATICAL CURVES

All mathematical curves may, in principle, be produced, to a given approximation, by a numerically controlled system. However, for a curve defined *a priori*, the straight line or the circular arc, or, rather, an assembly made up of segments of these curves, is almost always used. They are defined according to the method explained on p. 99.

The best known systems are Profiledata (Ferranti)[8], Paget (Olivetti), A.P.T., etc.

NON-MATHEMATICAL CURVES

To define an experimental curve, it is generally sufficient to measure the coordinates of a certain number of its points. Measurement of the directions of the tangents in addition is rather uncommon, as these are much more difficult to measure than a coordinate.

A certain number of machines (Pratt and Whitney, Cincinnati, Heyligenstaedt, Aristo, Ferranti, D.E.A., Schiess, etc.) are capable of making these readings in an automatic or semiautomatic manner (Figure 3.7).

When this information has been obtained, there are several ways of using it.

Definition of a single curve

Given $n+1$ points with coordinates x_i, $y_i (0 \leq i \leq n)$, there is a function:

$$y = \sum_{0}^{n} a_i \,.\, x^i \tag{5.1}$$

which is a curve satisfying the given conditions. The coefficients a_i are the solutions of a system of $n+1$ linear equations, each of which is obtained by giving, in succession, for the values of x and y, the coordinates of the $n+1$ points through which the curve must pass.

If the curve is allowed to pass within a certain distance of the given points, the method of least squares may sometimes be used to find a polynomial function of degree less than n satisfying the conditions imposed.

This method of representation, whatever the variant, has the advantage of lending itself easily to an automatic treatment of the data, but it has two serious drawbacks:

(a) A curve defined by equation 5.1 cannot have a vertical tangent. As a result, the points are arranged in order of increasing x.

This difficulty can be overcome by using successively two orthogonal-axis systems staggered by 90°. For example, in Figure 5.4, the segments AB and BC are, respectively, referred to the axes x_1Oy_1 and x_2Oy_2.

The change from one set of axes to the other may be an operator's decision; it can also be done automatically when the slope of the tangent reaches a certain limit, for example 50° in the increasing direction and 40° in the decreasing direction.

(b) The shape of the curve depends on the orientation of the coordinate axes. In effect, a group of points defines a certain curve; but if this group is rotated, the curve which it then defines is not the same as the previous curve rotated.

Thus, if the same shape is to be preserved, a curve must first be defined relative to the first coordinate axes, and then the rotation transformation

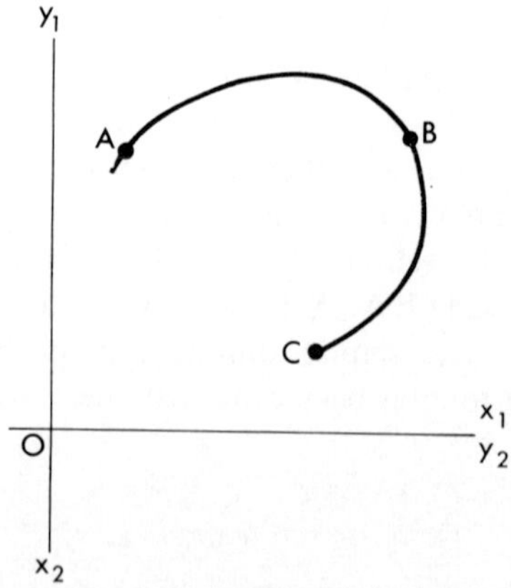

Fig. 5.4

must be applied to all the points on the curve between which interpolation is to be made; the consequence of this is to increase the number of calculations to be performed.

It is possible to transform a polynomial representation into parametric form, which is better suited to matrix methods.

Given a curve defined in a coordinate system by the relation

$$y = \sum_{0}^{n} a_i \,.\, x^i$$

let $\mathbf{u}_x$ and $\mathbf{u}_y$ be the unit vectors relative to the axes Ox and Oy. In a Cartesian system, the magnitudes of these vectors are identical and equal to the unit of measurement:

$$|\mathbf{u}_x| = |\mathbf{u}_y| = 1.$$

If v and a_i are, respectively, a parameter and scalars, we can write:

$$\mathbf{x}(v) = \mathbf{u}_x \,.\, v$$

$$\mathbf{y}(v) = \mathbf{u}_y \sum_{0}^{n} a_i \,.\, v^i$$

and

$$\mathbf{P}(v) = \mathbf{x}(v) + \mathbf{y}(v).$$

Piecewise definition

To find an expression satisfying simultaneously a large number of conditions is not a straightforward matter, so it is sometimes preferable to use a series of arcs, each of which joins two consecutive points through which the curve must pass and is tangential to the adjacent segments.

Arcs with predetermined tangents

Determination of the tangents

Certain rules exist that allow the definition of tangents; some examples are:

(a) The tangent at one point is parallel to the chord joining the adjacent points (Figure 5.5). The rule does not apply to the end points. In these cases, two imaginary points can be defined, one before the first point and the other after the last point.

It could also be arranged that the tangents to the first or last two points form an isosceles triangle with the chord which joins them.

(b) The tangent at one point is a tangent to the circle defined by this point and the two adjacent points. For the end points, one or other of the two preceding rules is used.

(c) Finally, it can be left to the designer to fix arbitrarily the direction of each tangent.

The first two methods can break down when the two successive arcs are of opposite curvature. It is seen (Figure 5.6) that the direction of the tangent at B is clearly different from that of the chord AC.

Some algorithms have been devised to detect these cases. For example (Figure 5.7), the intersections of the tangent at C, which is parallel to BD, with the straight lines through B and D and parallel, respectively, to AC and CE, are the points M and N, both situated on the same side of C, when normally the latter point is situated between them.

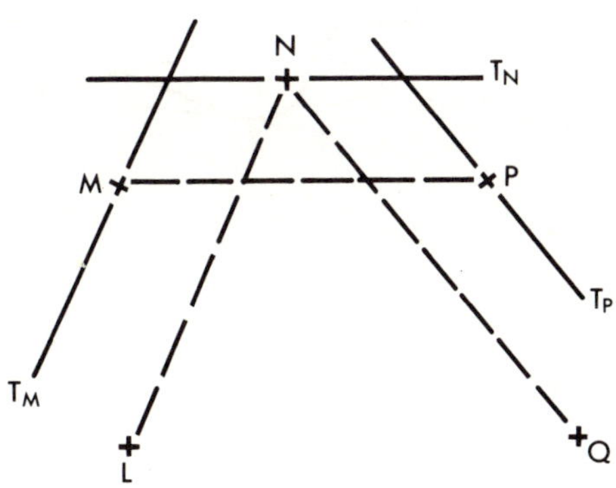

Fig. 5.5

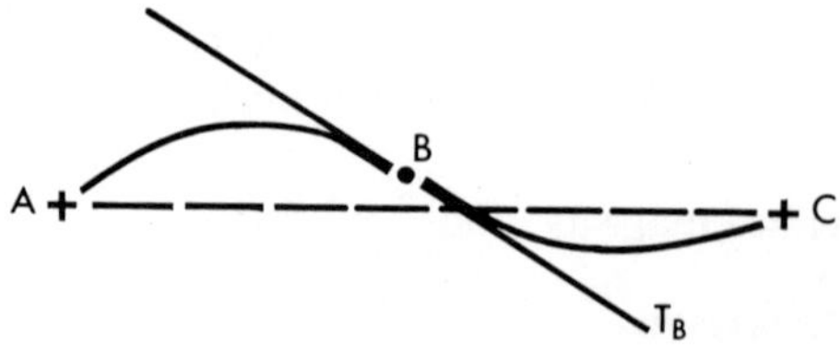

Fig. 5.6

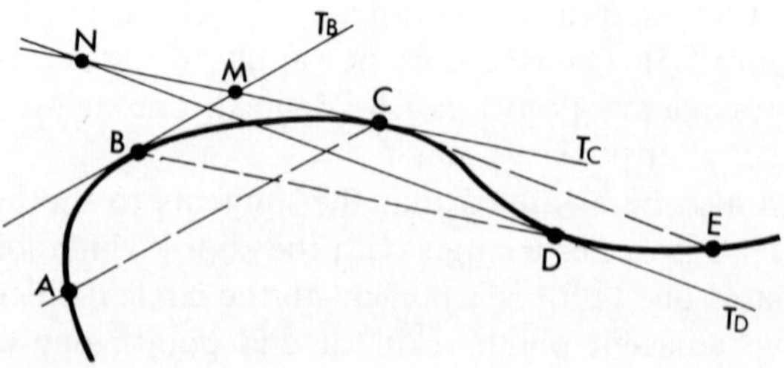

Fig. 5.7

If this is found to be the case, the tangent at C is given a direction which takes account of the directions of the tangents at B and D, as well as that of the chord BD.

This criterion is not universally correct, and can also be found in error (Figure 5.8). In this case, it is generally sufficient to give an arbitrary direction to the tangent, defined by a fictitious point as in the case of an end point.

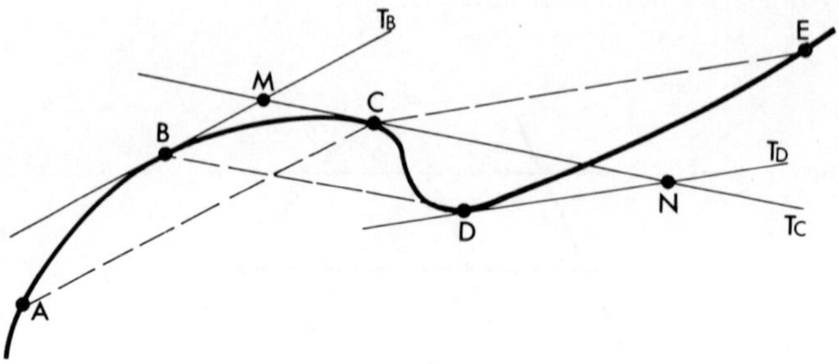

Fig. 5.8

Curves

For a segment defined by its end points and tangents, many curves solve this problem, and only the most commonly used solutions will be examined.

Arcs of circles. The segment is broken into two circular arcs (Figure 5.9). As the problem has an infinite number of solutions, one of the two radii r or R can be imposed, and the calculation program for the other can make use of one of the methods described on p. 99.

To treat this problem in a more automated manner, it is arranged that the chosen solution should correspond to the minimum value of the ratio $|R/r| = \lambda$.

From Figure 5.10, let

$$\mathrm{TM} = l$$

$$\mathrm{TN} = kl.$$

We see that:

$$|\mathrm{AN}| = kl \tan \alpha$$

$$|\mathrm{AT}| = \frac{kl}{\cos \alpha}$$

$$\mathrm{AM} = l\left(1+\frac{k}{\cos \alpha}\right) = l\frac{\cos \alpha + k}{\cos \alpha}$$

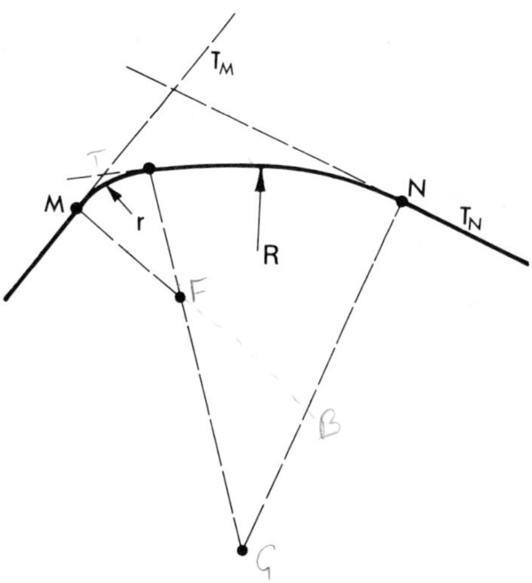

Fig. 5.9

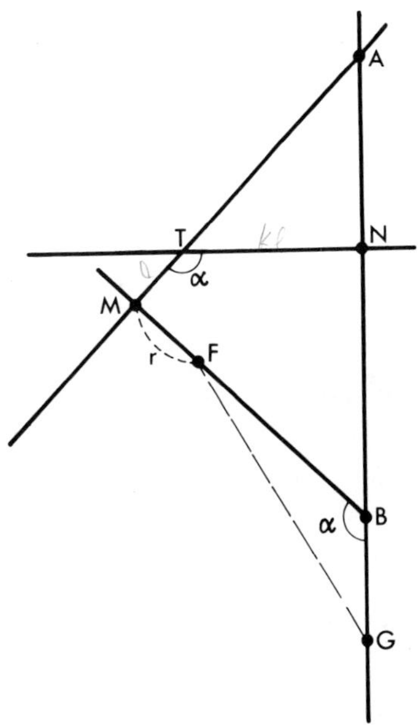

Fig. 5.10

$$\mathrm{MB} = \frac{\mathrm{AM}}{\tan\alpha} = l\frac{\cos\alpha + k}{\sin\alpha}$$

$$\mathrm{AB} = \frac{\mathrm{AM}}{\cos\alpha} = l\frac{\cos\alpha + k}{\sin\alpha\cos\alpha}$$

$$\mathrm{NB} = \mathrm{AB} - \mathrm{AN} = l\frac{\cos\alpha + k}{\sin\alpha\cos\alpha} - kl\frac{\sin^2\alpha}{\sin\alpha\cos\alpha}$$

$$= l\frac{\cos\alpha + k(1-\sin^2\alpha)}{\sin\alpha\cos\alpha} = l\frac{1+k\cos\alpha}{\sin\alpha}.$$

If we set $r = \mathrm{MF}$, we have the relationships:

$$\mathrm{FG} = \mathrm{GN} - \mathrm{FM} = (R-r)$$

$$\overline{\mathrm{FG}}^2 = \overline{\mathrm{GB}}^2 + \overline{\mathrm{BF}}^2 - 2\mathrm{GB}\,.\,\mathrm{BF}\,.\cos\alpha$$

$$(R-r)^2 = (R-\mathrm{BN})^2 + (\mathrm{BM}-r)^2 - 2(R-\mathrm{BN})(\mathrm{BM}-r)\,.\cos\alpha$$

$$R^2 - 2Rr + r^2 = R^2 - 2R\,.\,\mathrm{BN} + \mathrm{BN}^2 + \mathrm{BM}^2 - 2\mathrm{BM}\,.\,r + r^2$$
$$-2(R\,.\,\mathrm{BM} - \mathrm{BN}\,.\,\mathrm{BM} - Rr + \mathrm{BN}\,.\,r)\cos\alpha.$$

From which

$$-2Rr + 2R\,.\,\mathrm{BN} + 2R\,.\,\mathrm{BM}\,.\cos\alpha - 2Rr\cos\alpha$$
$$= R(-2r + 2\mathrm{BN} + 2\mathrm{BM}\cos\alpha - 2r\,.\cos\alpha)$$
$$= 2R[\mathrm{BN} + \mathrm{BM}\cos\alpha - r(1+\cos\alpha)]$$

$$\mathrm{BN}^2 + \mathrm{BM}^2 - 2\mathrm{BM}\,.\,r + 2\mathrm{BN}\,.\,\mathrm{BM}\cos\alpha - 2\mathrm{BN}\,.\,r\cos\alpha$$
$$= \mathrm{BN}^2 + \mathrm{BM}^2 + 2\mathrm{BN}\,.\,\mathrm{BM}\,.\cos\alpha - 2r(\mathrm{BM} + \mathrm{BN}\,.\cos\alpha)$$

$$R = \frac{\mathrm{BN}^2 + \mathrm{BM}^2 + 2\mathrm{BN}\,.\,\mathrm{BM}\,.\cos\alpha - 2r(\mathrm{BM} + \mathrm{BN}\cos\alpha)}{2[\mathrm{BN} + \mathrm{BM}\cos\alpha - r(1+\cos\alpha)]}.$$

For simplicity, set

$$\mathrm{BN}^2 + \mathrm{BM}^2 + 2\mathrm{BN}\,.\,\mathrm{BM}\,.\cos\alpha = a$$
$$-(\mathrm{BM} + \mathrm{BN}\,.\cos\alpha) = b$$
$$(\cos\alpha + 1) = c$$
$$\mathrm{BN} + \mathrm{BM}\,.\cos\alpha = d.$$

We then have:

$$\lambda = \frac{R}{r} = \frac{ar+b}{2(cr^2+dr)}.$$

This ratio is a minimum for $d\lambda/dr = 0$. Putting

$$u = ar+b, \qquad u' = a$$

$$v = cr^2+dr, \qquad v' = 2cr+d$$

$$\frac{d\lambda}{dr} = \frac{a(cr^2+dr)-(ar+b)(2cr+d)}{2(cr^2+dr)^2}$$

$$\frac{d\lambda}{dr} = 0 \quad \text{for} \quad a(cr^2+dr)-(ar+b)(2cr+d)$$

$$= acr^2+adr-2acr^2-2bcr-adr-bd$$

$$= -acr^2+(ad-2bc-ad)r-bd = 0.$$

If there are two real roots, they correspond to two solutions (Figures 5.11 and 5.12), of which one only is acceptable. Instead of performing all the preceding calculations, it seems much simpler to use a nomograph which, given the values of l, k and α, gives an approximate value for r.

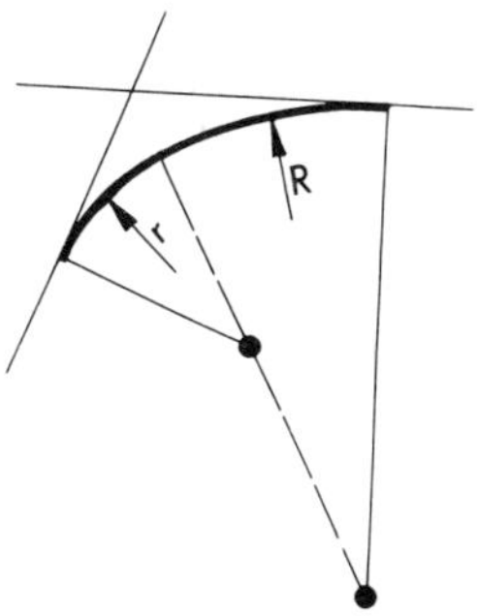

Fig. 5.11

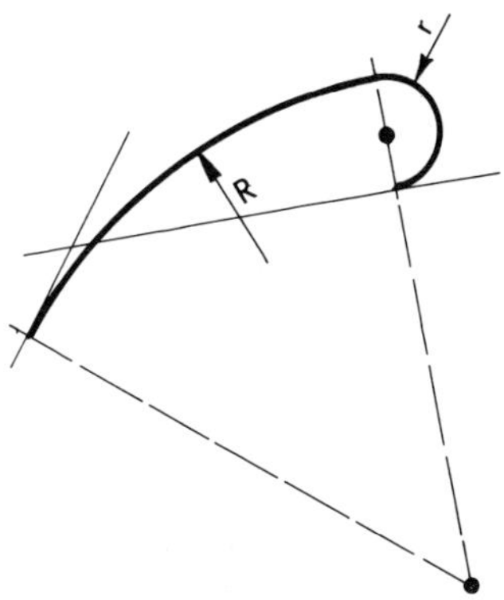

Fig. 5.12

Parabolic arcs. The three points M, N and T (Figure 5.13) define a parabola.

Cubic arcs. If parabolic arcs are used, it is essential that the changes in the direction of curvature should be located at the points common to two arcs.

To avoid this, curves defined by the equation

$$y = ax^3+bx^2+cx+d$$

are used, the coefficients being defined by the end conditions (Figure 5.14).

In contrast to the two preceding methods, this definition is not independent of the coordinate axes, since the axis of the curve is parallel to Oy.

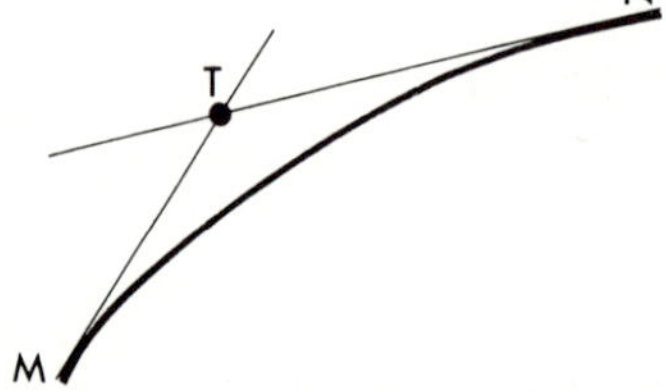

Fig. 5.13

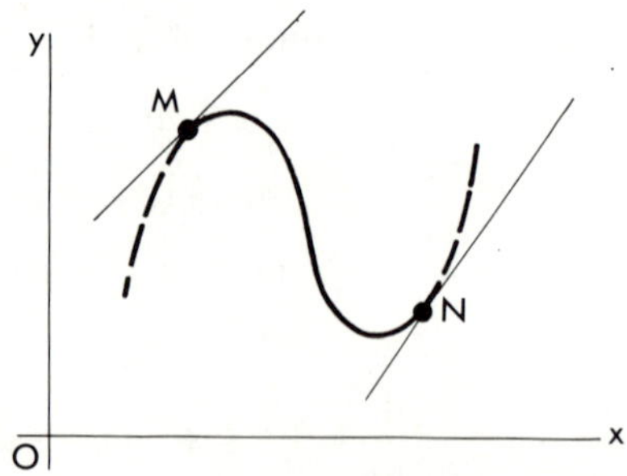

Fig. 5.14

To overcome this restriction, it can be arranged that each segment of the curve is referred to particular axes (Figure 5.15) whose origin is coincident with that of the segment and whose x axis passes through the end point of the segment.

However, these arcs still cannot have tangents parallel to the y axes, and, in the case of Figure 5.16, the solution would be represented by the line 1, whereas to obtain the curve 2, it would be necessary to make a preliminary division of this arc into two segments.

Segments determined by recurrence

The tangents are determined, step by step, by a certain number of constraints; some examples are given below.

Circular arcs

An arc AB is defined by its two end points and the tangent at A, which is identical with that of the preceding segment (Figure 5.17).

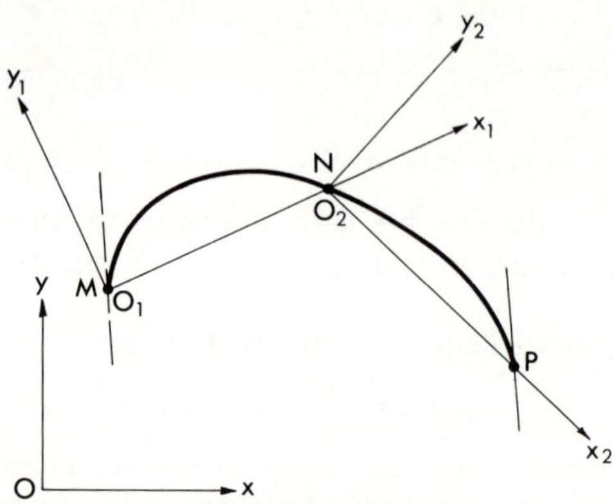

Fig. 5.15

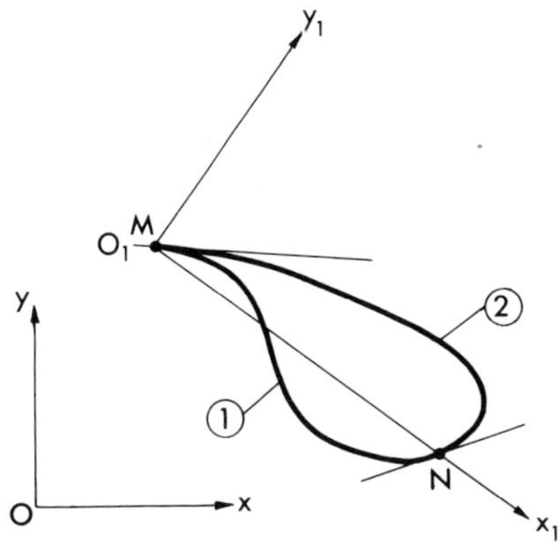

Fig. 5.16

The tangent at B is thus automatically defined, and serves as the tangent at the start of the arc BC.

As the piecewise curve is constrained to pass through a given number of points, the tangent at one of these points can be chosen arbitrarily.

Parabolic or cubic arcs

First-order continuity. A curve is defined by its initial point A and the tangent of the preceding arc as well as by the two following points B and C (Figure 5.18).

Only the arc AB, whose tangent at B, together with the points C and D, serves to define the arc BC, is retained. As with the process described on p. 105, this method leaves freedom of choice of one of the tangents, but it can, with conics, run into an impossibility (an imaginary solution) (Figure 5.19).

Other curves with the same properties have been proposed by different authors.[14]

If cubics are used, the case where the arc found passes successively through A, C and B and not through A, B and C can arise, and this case is a practical impossibility.

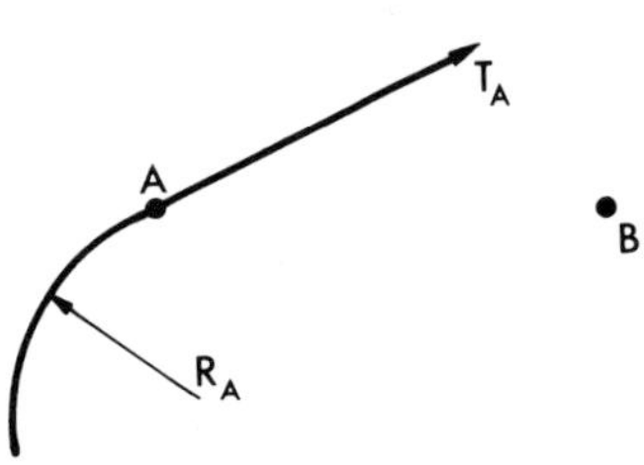

Fig. 5.17

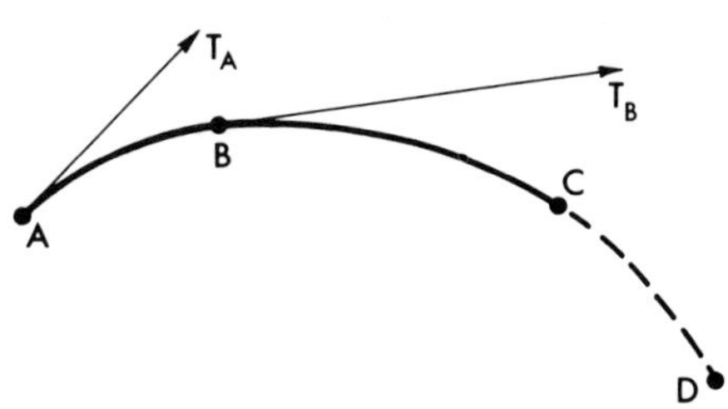

Fig. 5.18

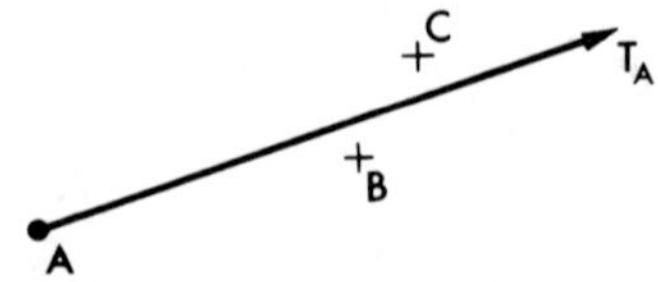

Fig. 5.19

Some difficulties of this kind are found in all the algorithms, since they are never suitable for solving all the cases.

Second-order continuity. In some problems, generally in fluid dynamics, there is an attempt to make two segments not only tangential but also osculatory.

The radius of curvature at a point depends only on the first and second derivatives.*

If the coordinates of the initial point A of a parabolic or cubic arc are given as well as the tangent and the curvature at this point, the arc is defined by the following point B (Figure 5.20).

Thus the tangent and the curvature at B are known, which, with the coordinates of C, suffice to determine, step by step, all the arcs making up the piecewise curve.

Instead of the tangent and curvature at the initial point being fixed arbitrarily, two values (slopes or curvatures) associated with any points on the piecewise curve may equally well be fixed. Normally the tangent at each extremity is fixed (see Appendix 1).

A problem arises when a piecewise curve includes a straight-line segment (Figure 5.21) as is the case in the waterlines of cargo-boat hulls.[15]

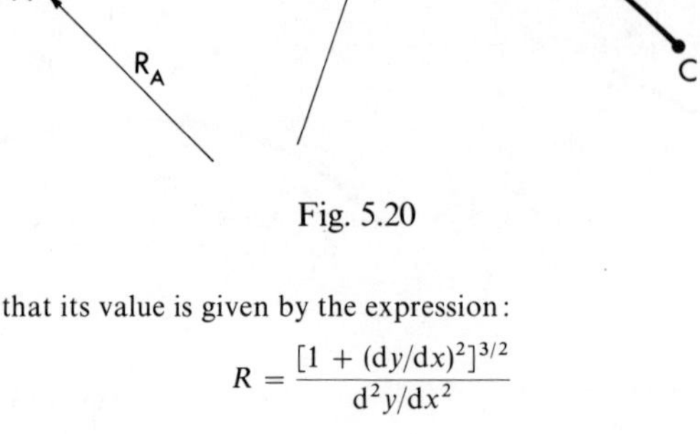

Fig. 5.20

* It is recalled that its value is given by the expression:

$$R = \frac{[1 + (dy/dx)^2]^{3/2}}{d^2y/dx^2}$$

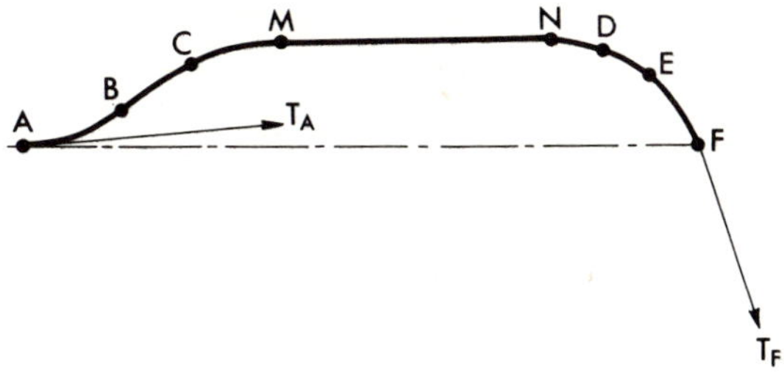

Fig. 5.21

Two conditions are imposed on M and N (tangent parallel to Ox and zero curvature), and the slope of the tangent at A and F cannot be fixed freely.

To retain this possibility, two of the arcs, CM and DE, for example, are subdivided into two segments which are second-order continuous at a point situated between C and M and between D and E.

Parametric curves

Ferguson's curves[16]

A point on an arc of a curve is represented by the expression

$$\mathbf{P}_{(v)} = \mathbf{m}v^3 + \mathbf{n}v^2 + \mathbf{r}v + \mathbf{q}^*$$

in which v is a parameter.

To define the four vectors **m**, . . . , **q**, the following conditions are imposed (Figure 5.22):

1. At the initial point A, the parameter v takes the value O.
2. At the final point B, the parameter takes the value 1.
3. The values of the coordinates at A and B, as well as those of the parametric derivative† of the curve at these points are given.

The vectors **m**, . . , **q**, are the solutions of the system of equations:

$$\mathbf{P}_{(0)} = \mathbf{A} = \mathbf{q}$$

$$\mathbf{P}_{(1)} = \mathbf{B} = \mathbf{m} + \mathbf{n} + \mathbf{p} + \mathbf{q}$$

$$\frac{d\mathbf{P}_{(0)}}{dv} = \frac{d\mathbf{A}}{dv} = \mathbf{p}$$

$$\frac{d\mathbf{P}_{(1)}}{dv} = \frac{d\mathbf{B}}{dv} = 3\mathbf{m} + 2\mathbf{n} + \mathbf{p}$$

* **P** is a vector with its origin at the origin of the coordinate system and extremity at the point P.

† Note that this derivative is represented by a vector which is a tangent to the curve and not simply by the direction of the tangent.

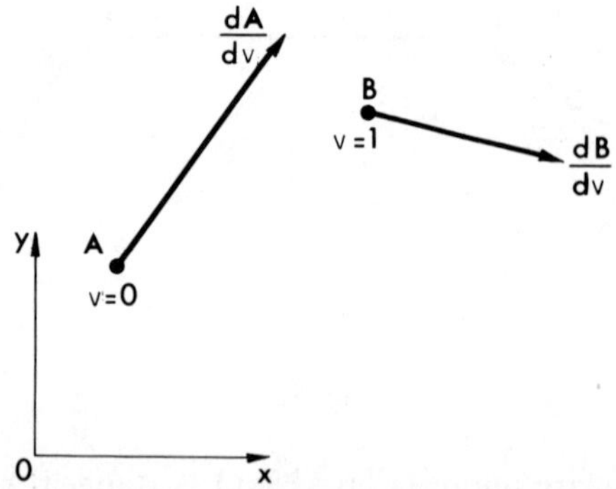

Fig. 5.22

From which we deduce:

$$\mathbf{m} = 2\left(\mathbf{A}-\mathbf{B}+\frac{d\mathbf{A}}{dv}\right)+\frac{d\mathbf{B}}{dv}$$

$$\mathbf{n} = 3\left(\mathbf{B}-\mathbf{A}-\frac{d\mathbf{A}}{dv}\right)-\frac{d\mathbf{B}}{dv}$$

$$\mathbf{p} = \frac{d\mathbf{A}}{dv}$$

$$\mathbf{q} = \mathbf{A}.$$

Among the properties of a curve segment defined in this way, the following are worth mentioning:

(a) A segment is completely determined by the quantities

$$\mathbf{A}, \mathbf{B}, \frac{d\mathbf{A}}{dv} \text{ and } \frac{d\mathbf{B}}{dv}.$$

(b) It has A for the initial point and B for the final point.
(c) The tangents at A and B are, respectively, coincident with the axes of the vectors $d\mathbf{A}/dv$ and $d\mathbf{B}/dv$.
(d) Curves having the same initial points and the same final points, and whose derivative vectors have the same directions but different magnitudes, are tangents at A and B, but their shape can be influenced by varying the magnitude of these vectors.
(e) The shape of the curve is independent of the coordinate system. Thus a rotation and translation transformation may be applied to the ensemble of points and vectors without changing the shape of the curve.

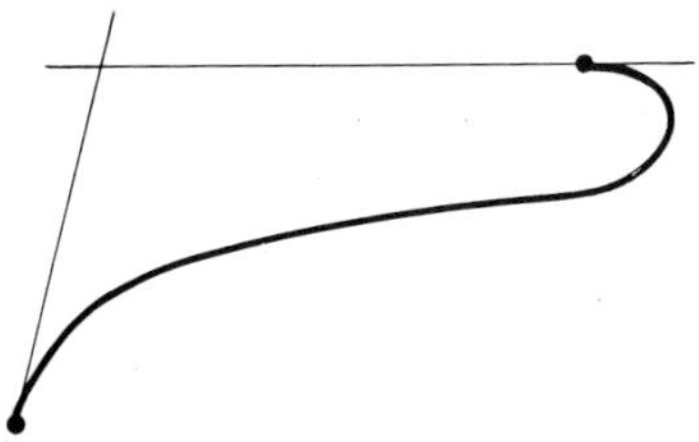

Fig. 5.23

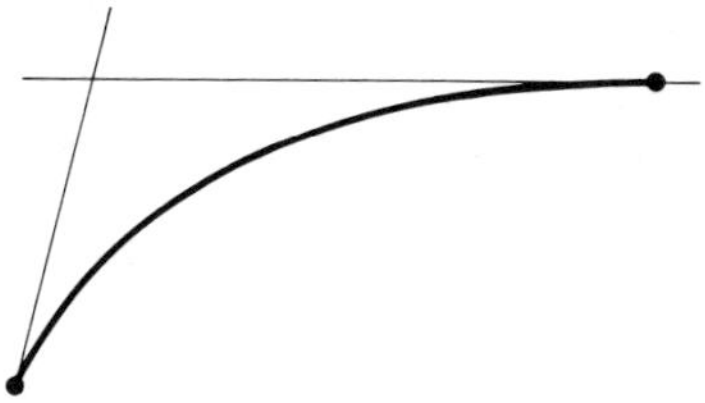

Fig. 5.24

(f) As the derivative vectors are directed, the cases evoked on p. 109 and p. 111 do not arise, i.e. a curve with the shape of Figure 5.23 or Figure 5.24 cannot be obtained if a curve such as that of Figure 5.25 is sought.

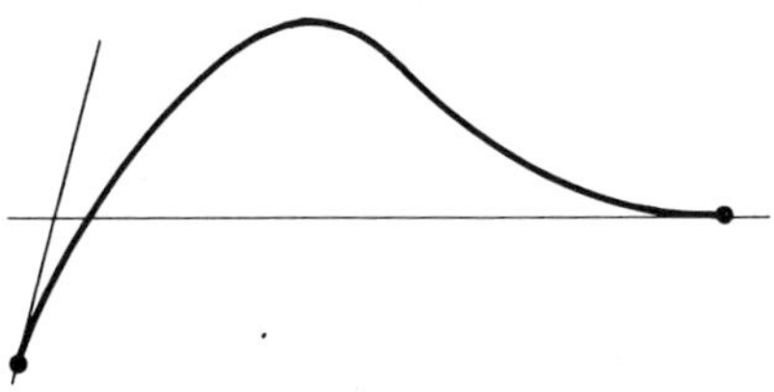

Fig. 5.25

(g) A segment can have two points of inflexion.

In effect, the radius of curvature is expressed by the formula

$$|R| = \frac{|\mathrm{d}\mathbf{P}/\mathrm{d}v|^3}{|\mathrm{d}\mathbf{P}/\mathrm{d}v \wedge \mathrm{d}^2\mathbf{P}/\mathrm{d}v^2|}$$

whose denominator can have two roots when v ranges from 0 to 1 without $\mathrm{d}\mathbf{P}/\mathrm{d}v$ itself being zero.

UNISURF curves

The UNISURF system is studied principally because of its application to the operations of drawing automobile bodies and milling the corresponding stamping tools, but it can be used to solve other problems.[17]

Principle. A point on a curve segment is represented, using the same notation as in the preceding section, by the expression:

$$\mathbf{P}_{(v)} = \sum_{0}^{n} \mathbf{a}_i \cdot f_i(v). \tag{5.2}$$

Properties of the functions f. A great variety of f functions can be used, according to the proposed objectives. Those which are currently used are defined by the following conditions:

(a) The initial point of a segment corresponds to the value 0 of the parameter, and we have:

$$\mathbf{P}_{(0)} \equiv \mathbf{a}_0.$$

Thus

$$f_0(0) \equiv 1$$

and

$$f_i(0) \equiv 0 \qquad \text{for} \quad 1 \leq i \leq n.$$

(b) The final point of a segment corresponds to the value 1 of the parameter and we have:

$$\mathbf{P}_{(1)} = \sum_0^n \mathbf{a}_i.$$

Thus:

$$f_i(1) \equiv 1 \quad \text{for} \quad 0 \leq i \leq n.$$

We have, further:

$$f_0(v) \equiv 1.$$

As a result, equation 5.2 may be written:

$$\mathbf{P}_{(v)} = \mathbf{a}_0 + \sum_1^n \mathbf{a}_i \,.\, f_i(v). \tag{5.3}$$

(c) The tangent at the initial point is parallel to the vector $\mathbf{a}_1$. Now, we have:

$$\mathbf{P}'_{(v)} = \sum_0^n \mathbf{a}_i \,.\, f_i'(v). \tag{5.4}$$

but since $f_0(v)$ = constant

$$\mathbf{P}'_{(v)} = \sum_1^n \mathbf{a}_i \,.\, f_i'(v).$$

In order that:

$$\mathbf{P}_{(0)} = k \,.\, \mathbf{a}_1 \quad (\text{with } k > 0)$$

it follows that

$$f'_1(0) > 0 \quad \text{and} \quad f_i'(0) = 0 \quad \text{for} \quad 2 \leq i \leq n.$$

(d) The tangent at the final point is parallel to the vector $\mathbf{a}_n$. By analogy, with the conditions (c) above, it follows that:

$$f'_n(1) > 0 \quad \text{and} \quad f'_i(1) = 0 \quad \text{for} \quad 1 \leq i \leq n-1.$$

(e) The osculating plane at the origin is parallel to the vectors $\mathbf{a}_1$ and $\mathbf{a}_2$. It is known that the binormal at a point on the curve is defined by:

$$\mathbf{B}_{(v)} = \mathbf{P}'_{(v)} \wedge \mathbf{P}''_{(v)}.$$

In order that its direction at the point $\mathbf{P}_{(0)}$ should depend only on $\mathbf{a}_1$ and $\mathbf{a}_2$, we must have:

$$f''_1(0) \neq 0 \neq f''_2(0) \quad \text{and} \quad f''_i(0) \equiv 0 \quad \text{for} \quad 3 \leq i \leq n.$$

(f) The osculating plane at the final point is parallel to the vectors $\mathbf{a}_{n-1}$ and $\mathbf{a}_n$. By analogous reasoning, we deduce that we must have:

$$f''_n(1) \neq 0 \neq f''_{n-1}(1) \quad \text{and} \quad f''_i(1) = 0 \quad \text{for} \quad 1 \leq i \leq n-2.$$

(g) The family of the functions f is symmetric with regard to v and $(1-v)$. Thus we have:

$$f_i(v) \equiv 1 - f_{(n-i+1)}(1-v).$$

As a result of this property, it is possible to permute the initial and final points of a curve without changing its shape. It is sufficient to reverse the order of the indices of the vectors and to reverse their directions.

(h) If the number n of the functions is greater than 3, we deduce, by extrapolation, properties analogous to those of (c), (d), (e) and (f) above, i.e.:

$$f'''_1(0), \ldots, f'''_3(0) \neq 0 \quad \text{and} \quad f'''_4(0) \equiv \cdots \equiv f'''_n(0) \equiv 0$$

$$f''''_1(0), \ldots, f''''_4(0) \neq 0 \quad \text{and} \quad f''''_5(0) \equiv \cdots \equiv f''''_n(0) \equiv 0$$

....

$$f'''_n(1), \ldots, f'''_{n-2}(1) \neq 0 \quad \text{and} \quad f'''_{n-3}(1) \equiv \cdots \equiv f'''_1(1) \equiv 0$$

$$f''''_n(1), \ldots, f''''_{n-3}(1) \neq 0 \quad \text{and} \quad f''''_{n-4}(1) \equiv \cdots \equiv f''''_1(1) \equiv 0$$

....

From the properties described in (e), (f) and (h), if the vector $\mathbf{a}_1$ or the vector $\mathbf{a}_n$ is zero, the curve is a tangent to the vector $\mathbf{a}_2$ or $\mathbf{a}_{n-1}$, and so on.

It is thus a tangent, at its end point, to the first and last non-zero vectors.

Form of the functions f. Among all the families of functions satisfying the conditions (a) to (h) above, one of the simplest to use is of the form:

$$\underset{n}{f_i}(v) = \frac{(-v)^i}{(i-1)!} \cdot \frac{d^{i-1}\phi_n}{dv^{i-1}} \tag{5.5}$$

with

$$\phi_n = \frac{(1-v)^n - 1}{v}. \tag{5.5}$$

For example, setting $n = 5$, the following family is obtained:

$$\underset{5}{f_1} = v^5 - 5v^4 + 10v^3 - 10v^2 + 5v$$

$$\underset{5}{f_2} = -4v^5 + 15v^4 - 20v^3 + 10v^2$$

$$\underset{5}{f_3} = 6v^5 - 15v^4 + 10v^3$$

$$\underset{5}{f_4} = -4v^5 + 5v^4$$

$$\underset{5}{f_5} = v^5.$$

The application of this formula verifies that $\underset{n}{f_{n+1}}$ is identically zero.

If we assume that

$$\frac{d^0\phi}{dv^0} = \phi$$

we indeed obtain the function $\underset{n}{f_1}$.

But if we admit that

$$\frac{d^{-1}\phi}{dv^{-1}} = \int \phi \, dv$$

we find $\underset{n}{f_0} \neq 1$.

However, notice that the functions $\underset{n}{f_i}$ are expressed in the form

$$\underset{n}{f_i} = (-1)^{p+1} \sum_{p=0}^{n} \sum_{i=0}^{n} C_{p-1}^{p-i} C_n^p v^p.$$

It is evident that

$$C_l^k \equiv 0 \quad \text{if} \quad k < l.$$

Moreover, if we allow that $C_l^0 \equiv 1$ since

$$C_l^k = \frac{l!}{(l-k)!k!}$$

and for $k = 0$

$$C_l^0 = \frac{l!}{(l-0)!0!} = \frac{l!}{l!0!} = \frac{1}{0!} = 1,$$

we thus have

$$\underset{n}{f_0}(v) \equiv 1$$

We have, in effect, for the case $n = 3$

$$\underset{3}{f_0} = (-1)^0(C^0_{-1} \,.\, C^0_3 \,.\, v^0 - C^1_0 \,.\, C^1_3 \,.\, v + C^2_1 \,.\, C^2_3 \,.\, v^2 - C^3_2 \,.\, C^3_3 \,.\, v^3)$$

with

$$C^0_{-1} = C^0_3 = 1 \quad \text{and} \quad C^1_0 = C^2_1 = C^3_2 = 0$$

therefore

$$\underset{3}{f_0} = 1.$$

By arranging the vectors $\mathbf{a}_0, \mathbf{a}_1, \ldots, \mathbf{a}_n$ end to end, a polygon called the characteristic polygon is constructed, whose vertices S_0 and S_n coincide with the ends of the curve segment being defined (Figure 5.26).

This representation is convenient because the shape of the polygon $S_0, \ldots, S_n$ enables the shape of the curve C to be imagined easily, given some elementary knowledge of geometry.

To perform certain calculations, it is more simple to use the vectors $\mathbf{S}_1, \ldots, \mathbf{S}_n$ directly, so that:

$$\begin{aligned} \mathbf{S}_0 &= \mathbf{a}_0 \\ \mathbf{S}_1 &= \mathbf{a}_0 + \mathbf{a}_1 \\ \mathbf{S}_2 &= \mathbf{a}_0 + \mathbf{a}_1 + \mathbf{a}_2 \\ \mathbf{S}_3 &= \mathbf{a}_0 + \mathbf{a}_1 + \mathbf{a}_2 + \mathbf{a}_3 \end{aligned} \tag{5.6}$$

or in a more general form

$$\mathbf{S}_i = \sum_0^i \mathbf{a}_i. \tag{5.7}$$

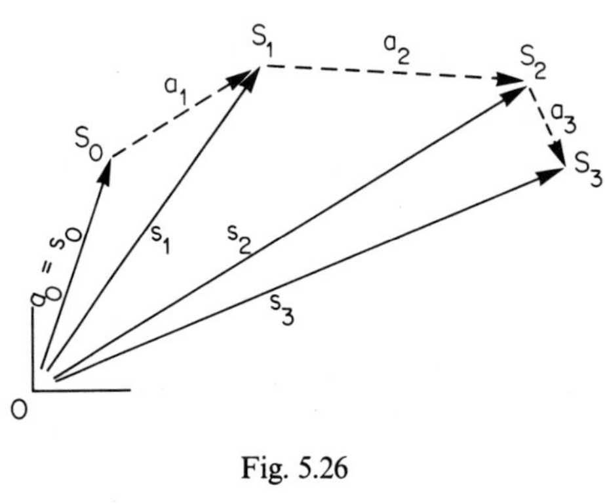

Fig. 5.26

In these circumstances, the point $\mathbf{P}_{(v)}$ on the curve is represented by the expression

$$\mathbf{P}_{(v)} = \sum_{0}^{n} \mathbf{S}_i \cdot \underset{n}{g_i}(v).$$

To derive the functions $\underset{n}{g_i}$ from the functions $\underset{n}{f_i}$, in the case where $n = 3$, we set:

$$\begin{aligned} \mathbf{P}_{(v)} &= \mathbf{a}_0+\mathbf{a}_1(v^3-3v^2+3v)+\mathbf{a}_2(-2v^3+3v^2)+\mathbf{a}_3 \cdot v^3 \\ &= \mathbf{S}_0(cv^3+dv^2+ev+h)+\mathbf{S}_1(kv^3+lv^2+mv+p) \\ &\quad +\mathbf{S}_2(qv^3+rv^2+tv+u)+\mathbf{S}_3(wv^3+xv^2+yv+z). \end{aligned} \tag{5.8}$$

Taking equation 5.6 into account, we have:

$$\begin{aligned} \mathbf{P}_{(v)} = \mathbf{a}_0&[(c+k+q+w)v^3+(d+l+r+x)v^2+(e+m+t+y)v+(h+p+u+z)] \\ &+\mathbf{a}_1[(k+q+w)v^3+(l+r+x)v^2+(m+t+y)v+(p+u+z)] \\ &+\mathbf{a}_2[(q+w)v^3+(r+x)v^2+(t+y)v+(u+z)] \\ &+\mathbf{a}_3[wv^3+xv^2+yv+z]. \end{aligned} \tag{5.9}$$

By equating the coefficients of the $\mathbf{a}_i$ in equations 5.8 and 5.9, we have a linear system with sixteen unknowns whose roots are

$$\begin{array}{llll} c = -1 & k = 3 & q = -3 & w = 1 \\ d = 3 & l = -6 & r = 3 & x = 0 \\ e = -3 & m = 3 & t = 0 & y = 0 \\ h = 1 & p = 0 & u = 0 & z = 0. \end{array}$$

We thus have:

$$\begin{aligned} \mathbf{P}_{(v)} = \mathbf{S}_0&(-v^3+3v^2-3v+1) \\ &+\mathbf{S}_1(3v^3-6v^2+3v) \\ &+\mathbf{S}_2(-3v^3+3v^2) \\ &+\mathbf{S}_3v^3 \end{aligned}$$

where

$$\mathbf{P}_{(v)} = \mathbf{S}_0(1-v)^3+\mathbf{S}_1 \cdot 3v(1-v)^2+\mathbf{S}_2 \cdot 3v^2(1-v)+\mathbf{S}_3v^3.$$

The g functions are given by the general formula

$$\underset{n}{g_i} = \mathrm{C}_n^i \cdot v^i(1-v)^{n-i}$$

or again by

$$g_{\substack{i\\n}} = (-1)^{p+i} \sum_{i=0}^{n} \sum_{p=0}^{n} C_p^i C_n^p v^p$$

it being understood that $C_m^0 \equiv 1$ for all m and

$$C_l^k \equiv 0 \quad \text{if} \quad k > l.$$

To facilitate the computations, a representation analogous to that of Ferguson may equally well be used, i.e.

$$\mathbf{P}_{(v)} = \sum_0^n \mathbf{b}_i \,.\, v^i$$

or in a form which lends itself better to computation

$$\mathbf{P}_{(v)} = \{\ldots[(\mathbf{b}_n \,.\, v) + \mathbf{b}_{n-1}]v + \mathbf{b}_{n-2} \ldots\}v + \mathbf{b}_0 .$$

The vectors $\mathbf{b}_i$ are derived from the vectors $\mathbf{a}_i$ by the transformation:

$$[\mathbf{b}_i] = [M_{\mathbf{a},\mathbf{b},n}] \times [\mathbf{a}_i].$$

The matrices $[M_{\mathbf{a},\mathbf{b},n}]$ have the general form:

$$[M_{\mathbf{a},\mathbf{b},n}] = [(-1)^{l+c} C_{l-2}^{l-c} \,.\, C_n^{l-1}] \tag{5.10}$$

l and c being, respectively, the rank of the row and the column of the general term.

For $n = 3$ we find:

$$[M_{\mathbf{a},\mathbf{b},3}] = \begin{bmatrix} 1 & 0 & 0 & 0 \\ 0 & 3 & 0 & 0 \\ 0 & -3 & 3 & 0 \\ 0 & 1 & -2 & 1 \end{bmatrix}.$$

As before, we understand that

$$C_k^0 \equiv 1 \quad \text{for all } k$$

and that

$$C_m^l \equiv 0 \quad \text{if} \quad l > m.$$

Equation 5.5 can be explained geometrically:

Consider a characteristic polygon (Figure 5.27) with vertices A, M, N, Q, . . . , B.

Construct some points A′, M′, N′, Q′, . . . , such that

$$\frac{AA'}{AM} = \frac{MM'}{MN} = \cdots = v$$

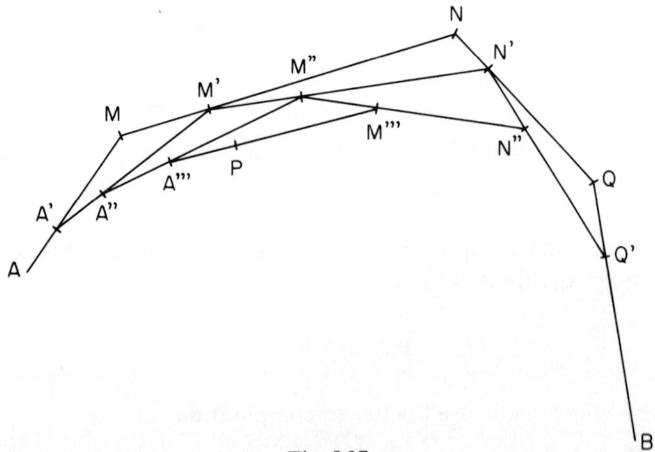

Fig. 5.27

and then some points A″, M″, N″ such that

$$\frac{A'A''}{A'M'} = \frac{M'M''}{M'N'} = \cdots = v$$

and so on.

On the final polygon, composed of the vector A″M″, there is a point P such that:

$$\frac{A''P}{A''M''} = v.$$

It may readily be proved that P is the point P(v) on the curve for which AMN ... B is the characteristic polygon and that the line A″M″ is the tangent at this point.

In the particular case where $n = 2$, we obtain the classical construction for the parabola.

Determination of the vectors $\mathbf{a}_i$. The vectors $\mathbf{a}_i$ are chosen on the basis of the following remarks: the curve defined by the characteristic polygon has, for its starting point, the starting point of the vector $\mathbf{a}_1$ and, for its end point, the end point of vector $\mathbf{a}_n$. At these points it is, respectively, tangent to the same vectors.

The intermediate vertices are deduced from the conditions imposed: a point on the curve and a tangent are represented, respectively, by the equations 5.3 and 5.4.

The ensemble of these equations forms a system in which the $\mathbf{a}_i$s and the v_is are the unknowns. It is linear with respect to the $\mathbf{a}_i$s and of degree n with respect to the v_is.

Supplementary conditions are imposed as follows:

$$v_0 = 0$$

and

$$v_n = 1.$$

Among the solutions of this system, only those for which the values of the v_is are arranged in the order of the indices are acceptable.

If supplementary conditions associated with the direction of the principal normals are imposed, the system becomes bilinear with respect to $\mathbf{a}_i$.

In fixing the values of radii, the degree of the $\mathbf{a}_i$s is raised, making the solution of the system laborious.

The simplest solution consists in:

1. Increasing the conditions given by points on the curve and tangents rather than by defining the centres of principal curvature.
2. Giving to the ith point among $n+1$ points $v_0, v_1, \ldots, v_i, \ldots, v_{n-1}, v_n$ the parametric value:

$$v_i = \frac{i}{n}.$$

In these conditions, the $\mathbf{a}_i$s are the roots of a linear system.

If second-order continuity between surfaces must be ensured, it is simpler to use Coons's methods (see p. 165).

However, in the particular case of a planar curve defined by a three-sided polygon, there is a simple method for tracing an arc whose ends are defined by

The coordinates.
The direction of the tangents.
The curvatures (in value and sign).

The vectors (Figure 5.28) being $\mathbf{a}_1$, $\mathbf{a}_2$ and $\mathbf{a}_3$, we have:

$$\mathbf{A}' = \mathbf{a}_1 \cdot f_1'(0)$$

$$\mathbf{A}'' = \mathbf{a}_1 \cdot f_1''(0) + \mathbf{a}_2 \cdot f_2''(0)$$

$$\mathbf{B}' = \mathbf{a}_3 \cdot f_3'(1)$$

$$\mathbf{B}'' = \mathbf{a}_2 \cdot f_2''(1) + \mathbf{a}_3 \cdot f_3''(1)$$

$$\rho_{\mathrm{A}} = \frac{|\mathbf{A}'|^3}{|\mathbf{A}' \wedge \mathbf{A}''|}$$

$$\rho_{\mathrm{B}} = \frac{|\mathbf{B}'|^3}{|\mathbf{B}' \wedge \mathbf{B}''|}.$$

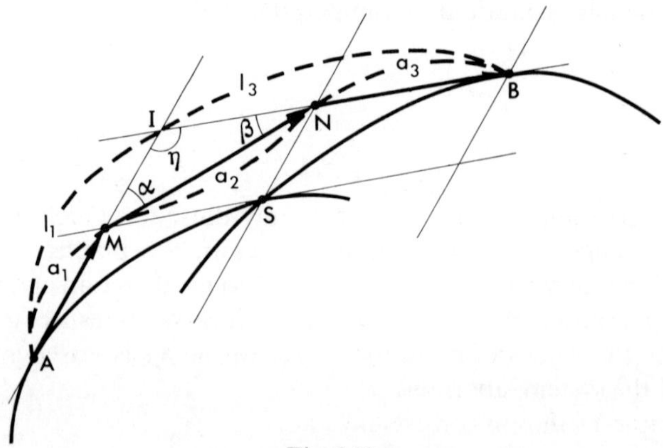

Fig. 5.28

Now:

$$\mathbf{A}' \wedge \mathbf{A}'' = \mathbf{a}_1 . f_1'(0) \wedge [\mathbf{a}_1 . f_1''(0) + \mathbf{a}_2 . f_2''(0)].$$

But:

$$\mathbf{a}_1 . f_1'(0) \wedge \mathbf{a}_1 . f_1''(0) \equiv 0.$$

Thus:

$$\begin{aligned} |\mathbf{A}' \wedge \mathbf{A}''| &= |\mathbf{a}_1 . f_1'(0) \wedge \mathbf{a}_2 . f_2''(0)| \\ &= |(\mathbf{a}_1 \wedge \mathbf{a}_2) . [f_1'(0) . f_2''(0)]| \\ &= |\mathbf{a}_1| \times |\mathbf{a}_2| . \sin\alpha . [f_1'(0) . f_2''(0)] \end{aligned}$$

and

$$\begin{aligned} \rho_A &= \frac{|\mathbf{a}_1|^3}{|\mathbf{a}_1| \times |\mathbf{a}_2|} . \frac{[f_1'(0)]^3}{[f_1'(0) \times f_2''(0) \sin\alpha]} \\ &= \frac{|\mathbf{a}_1|^2}{|\mathbf{a}_2|} \times \frac{[f_1'(0)]^2}{[f_2''(0) . \sin\alpha]}. \end{aligned}$$

Similarly:

$$\rho_B = \frac{|\mathbf{a}_3|^2}{|\mathbf{a}_2|} \times \frac{[f_3'(1)]^2}{[f_2''(1) . \sin\beta]}.$$

Now, in Figure 5.28, we see that:

$$\mathbf{a}_2 . \sin\alpha = \mathbf{MN} . \sin\alpha = \mathbf{IN} . \sin\eta$$

and

$$\mathbf{a}_2 . \sin\beta = \mathbf{MN} . \sin\beta = \mathbf{MI} . \sin\eta$$

whence:

$$\rho_A = \frac{|\overline{AM}|^2}{\overline{IN}} \cdot \frac{[f'_1(0)]^2}{[f''_2(0) \,.\, \sin \eta]}.$$

By setting:

$$AI = l_1$$
$$BI = l_3$$

we have:

$$\rho_A = \frac{|\mathbf{a}_1|^2}{|\mathbf{l}_3 - \mathbf{a}_3|} \cdot \frac{[f'_1(0)]^2}{f''_2(0) \,.\, \sin \eta} \tag{5.11}$$

and:

$$\rho_B = \frac{|\mathbf{a}_3|^2}{|\mathbf{l}_1 - \mathbf{a}_1|} \cdot \frac{[f'_3(1)]^2}{f''_2(1) \,.\, \sin \eta}. \tag{5.12}$$

By setting:

$$k = \frac{[f'_1(0)]^2}{f''_2(0) \,.\, \sin \eta} = \frac{[f'_3(1)]^2}{f''_2(1) \,.\, \sin \eta}.$$

equations 5.11 and 5.12 become:

$$\begin{aligned} \rho_A &= k \,.\, \frac{|\mathbf{a}_1|^2}{|\mathbf{l}_3 - \mathbf{a}_3|} \\ \rho_B &= k \,.\, \frac{|\mathbf{a}_3|^2}{|\mathbf{l}_1 - \mathbf{a}_1|}. \end{aligned} \tag{5.13}$$

From equations 5.13, we extract:

$$\mathbf{a}_3 = \mathbf{l}_3 - \frac{k}{\rho_A}|\mathbf{a}_1|^2. \tag{5.14}$$

Putting this value in equation 5.13, we obtain:

$$\frac{\rho_B}{k}|\mathbf{l}_1 - \mathbf{a}_1| = \left[\mathbf{l}_3 - \frac{k}{\rho_A}|\mathbf{a}_1|^2\right]^2$$

or, rearranging:

$$\left(\frac{k}{\rho_A}\right)^2 |\mathbf{a}_1|^4 - \frac{2k|\mathbf{l}_3|}{\rho_A}|\mathbf{a}_1|^2 + \frac{\rho_B}{k}|\mathbf{a}_1| + |\mathbf{l}_3|^2 - \frac{\rho_B}{k}|\mathbf{l}_1| = 0.$$

This equation has four real or imaginary roots.

If an approximate solution suffices, a chart based on the following principle is used:

The point S (Figure 5.28) is such that:

$$MS = (AM)^2 \cdot \frac{k}{\rho_A} \quad \text{and} \quad NS = (BN)^2 \cdot \frac{k}{\rho_B}.$$

It is thus at the intersection of two parabolas which, respectively, pass through A and B, have AI and BI for tangents, and have axes parallel to BI and AI.

To set up the chart, a transformation is used such that the transformed lines A_1I_1 and B_1I_1 are orthogonal and equal to unity (Figure 5.29).

In this new reference system, where the point S_1 has x and y coordinates transformed from $|\mathbf{l}_1 - \mathbf{a}_1|$ and from $|\mathbf{l}_3 - \mathbf{a}_3|$, equations 5.12 and 5.13 become:

$$\rho_A = \frac{k(1-x)^2}{y} \quad \text{where} \quad y = \frac{k}{\rho_A}(1-x)^2 \tag{5.15}$$

$$\rho_B = \frac{k(1-y)^2}{x} \quad \text{where} \quad x = \frac{k}{\rho_B}(1-y)^2. \tag{5.16}$$

The functions 5.15 and 5.16 are represented by the intersections of the parabolas whose coefficients depend, respectively, on ρ_A, ρ_B and η.

To define them (Figure 5.29), it seems convenient to characterize them by the points T, U, V and W where they cut the straight lines

$$y = \pm 1$$

and

$$x = \pm 1.$$

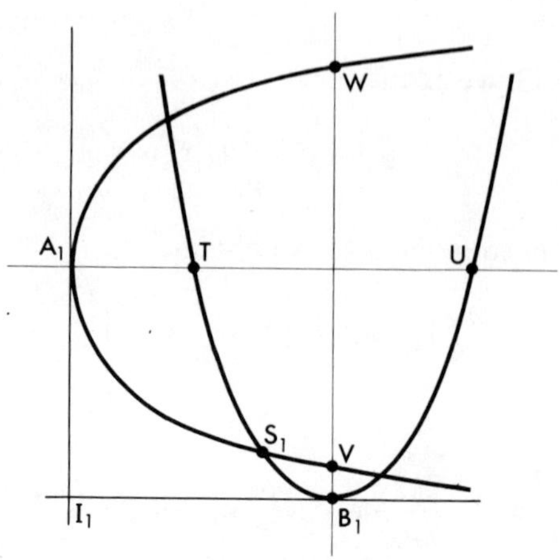

Fig. 5.29

We then have:

$$x_T = 1-\sqrt{\frac{\rho_A}{k}}$$

$$x_U = 1+\sqrt{\frac{\rho_A}{k}}$$

$$y_V = 1-\sqrt{\frac{\rho_B}{k}}$$

$$y_W = 1+\sqrt{\frac{\rho_B}{k}}.$$

Figure 5.30 readily reveals the existence and the value of the real solutions.

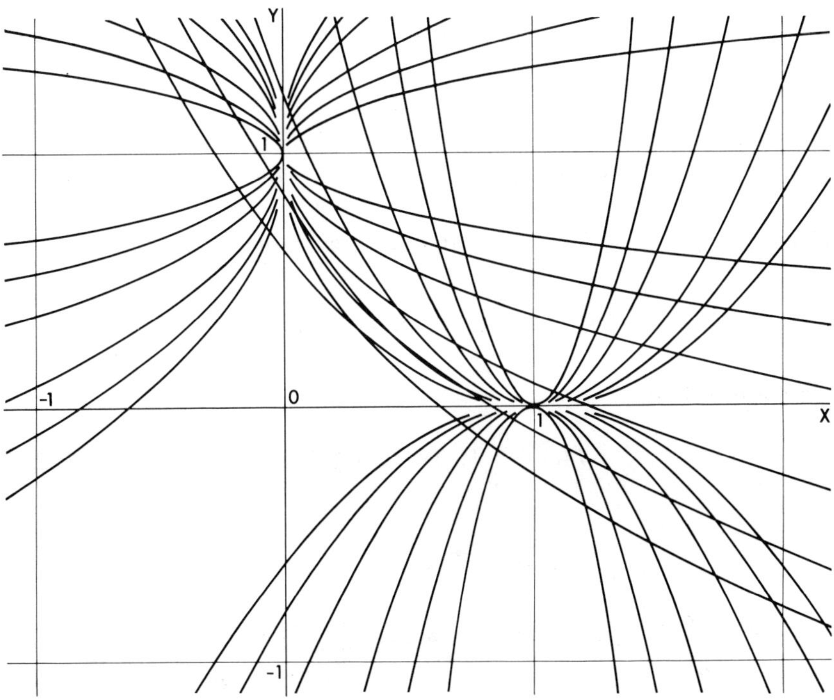

Fig. 5.30

If S_1 is one of the points of intersection, with coordinates x_{S_1} and y_{S_1}, the vectors $\mathbf{a}_1$ and $\mathbf{a}_3$ corresponding to this solution have as magnitudes:

$$|\mathbf{a}_1| = |\mathbf{l}_1| \,.\, (1-x_{S_1})$$

$$|\mathbf{a}_3| = |\mathbf{l}_3| \,.\, (1-y_{S_1}).$$

Hodographs. Consider a curve L defined by a characteristic polygon $\mathbf{a}_1, \mathbf{a}_2, \ldots, \mathbf{a}_n$ (Figure 5.31).

Let us draw (Figure 5.32) from the same origin O vectors equal in magnitude and direction to these vectors.

The polygon which joins their extremities defines a curve H.

Consider two points P(v) and Q(v) defined by the same parametric value, and situated, respectively, on L and H.

They possess the following property:

$$n \,.\, \mathbf{Q}(v) = \frac{d\mathbf{P}}{dr}(v). \tag{5.17}$$

H is thus the hodograph of L taking into account a factor of $1/n$.

In effect:

$$\frac{d\mathbf{P}}{dv} = \sum_{1}^{n} \mathbf{a}_i \,.\, f'_{\substack{i\\n}}(v) \tag{5.18}$$

and

$$\mathbf{Q}(v) = \mathbf{a}_1 + \sum_{1}^{n} (\mathbf{a}_{i+1} - \mathbf{a}_i) f_{\substack{i\\n-1}}(v) \tag{5.19}$$

and since $f_{\substack{0\\n}}(v) \equiv 1$, equation 5.19 can be written in the form:

$$\mathbf{Q}(v) = \sum_{1}^{n} \mathbf{a}_i \{ f_{\substack{i-1\\n-1}}(v) - f_{\substack{i\\n-1}}(v) \}. \tag{5.20}$$

We know that

$$f_{\substack{i\\n}}(v) = \sum_{0}^{n} (-1)^{i+p} C_n^p C_{p-1}^{p-i} v^p. \tag{5.21}$$

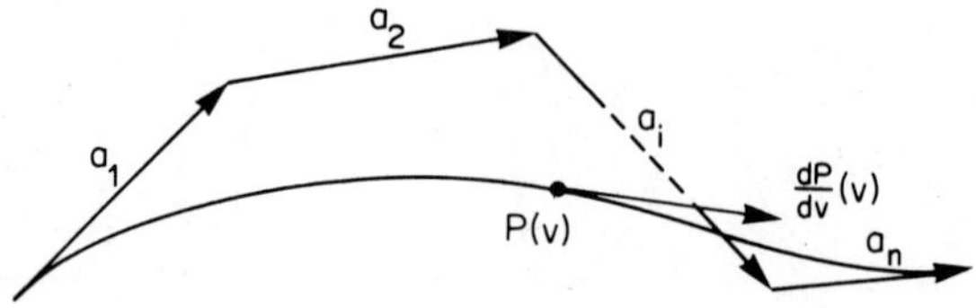

Fig. 5.31

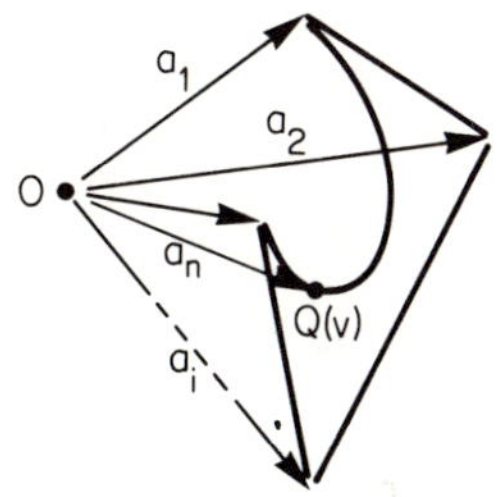

Fig. 5.32

The coefficients of rank lower than i are zero, since

$$C_{p-1}^{p-i} = 0 \quad \text{if} \quad p-i < 0.$$

Thus equation 5.21 is equivalent to

$$\underset{n}{f_i}(v) = \sum_{i}^{n} (-1)^{i+p} C_n^p C_{p-1}^{p-i} v^p \tag{5.22}$$

similarly

$$\underset{n-1}{f_i}(v) = \sum_{i}^{n-1} (-1)^{i+p} C_{n-1}^p C_{p-1}^{p-i} v^p$$

and

$$\underset{n-1}{f_{i-1}}(v) = \sum_{i-1}^{n-1} (-1)^{i+p-1} C_{n-1}^p C_{p-1}^{p-i+1} v^p. \tag{5.23}$$

In equation 5.23, let us extract from the sum the term corresponding to $p = i-1$ and let us raise by one the lower limit of the summation.

We obtain:

$$\underset{n-1}{f_{i-1}}(v) = (-1)^{2(i-1)} C_{n-1}^0 v^{i-1} + \sum_{i}^{n-1} (-1)^{i+p-1} C_{n-1}^p C_{p-1}^{p-i+1} v^p. \tag{5.24}$$

Since $C_k^0 \equiv 1$, we have:

$$\underset{n-1}{f_{i-1}}(v) = C_{n-1}^{i-1} v^{p-1} + \sum_{i}^{n-1} (-1)^{i+p-1} C_{n-1}^p C_{p-1}^{p-i+1} v^p. \tag{5.25}$$

Since $(-1)^{i+p} = -(-1)^{i+p-1}$, we have:

$$\underset{n-1}{f_{i-1}}(v) - \underset{n-1}{f_i}(v) = C_{n-1}^{i-1} v^{p-1} + \sum_{i}^{n-1} (-1)^{i+p-1} C_{n-1}^p v^p (C_{p-1}^{p-i+1} + C_{p-1}^{p-i}).$$

Now:

$$C_{p-1}^{p-i+1}+C_{p-1}^{p-i} = \frac{(p-1)!}{(p-i+1)!(i-2)!}+\frac{(p-1)!}{(p-i)!(i-1)!}$$

$$= (p-1)!\frac{(p-i)!(i-2)![(i-1)+(p-i+1)]}{(p-i)!(p-i+1)!(i-1)!(i-2)!}$$

$$= (p-1)!\frac{(p-i)!(i-2)!p}{(p-i)!(p-i+1)!(i-1)!(i-2)!}$$

$$= \frac{p(p-1)!}{(p-i+1)!(i-1)!} = \frac{p!}{(p-i+1)!(i-1)!} = C_p^{p-i+1}.$$

Therefore:

$$f_{\substack{i-1\\n-1}}(v)-f_{\substack{i\\n-1}}(v) = C_{n-1}^{i-1}v^{i-1}+\sum_{i}^{n-1}(-1)^{i+p-1}C_{n-1}^{p}C_{p}^{p-i+1}v^{p}.$$

Taking into account that $C_{n-1}^{i-1}v^{i-1} = (-1)^{i-p+1}C_{n-1}^{p}C_{p-1}^{p-i+1}v^{p}$ in the particular case where $p = i-1$, we can reintroduce this term in the expression under Σ, replacing the lower limit of i by $i-1$.

Thus:

$$f_{\substack{i-1\\n-1}}(v)-f_{\substack{i\\n-1}}(v) = \sum_{i-1}^{n-1}(-1)^{i-p+1}C_{n-1}^{p}C_{p-1}^{p-i+1}v^{p}. \tag{5.26}$$

Differentiating equation 5.22, we obtain:

$$f'_{\substack{i\\n}}(v) = \sum_{i}^{n}(-1)^{i+p}C_n^pC_{p-1}^{p-i}pv^{p-1}$$

$$= \sum_{i-1}^{n-1}(-1)^{i+p+1}C_n^{p+1}C_p^{p-i+1}(p+1)v^p.$$

Now:

$$(p+1)C_n^{p+1} = pC_{n-1}^{p}.$$

Because

$$(p+1)C_n^{p+1} = (p+1)\frac{n!}{(p+1)!(n-p-1)!} = \frac{n!}{p!(n-p-1)!}$$

and

$$nC_{n-1}^{p} = \frac{n.(n-1)!}{p!(n-p-1)!} = \frac{n!}{p!(n-p-1)!}.$$

Thus:

$$\frac{\underset{n}{f'_i}(v)}{\underset{n-1}{f_{i-1}}(v)-\underset{n-1}{f_i}(v)} = n\,\frac{\sum\limits_{i-1}^{n-1}(-1)^{i+p+1}C_p^{p-i+1}C_{n-1}^{p}}{\sum\limits_{i-1}^{n-1}(-1)^{i+p-1}C_p^{p-i+1}C_{n-1}^{p}} = n.$$

If a curve L and its hodograph H are projected on the same plane, their projections (l) and (h) have the following properties:

1. If $\overline{\mathrm{OQ}}(v)$ is tangent to (h), P(v) is a point of inflexion of (l).
2. If $\overline{\mathrm{OQ}}(v)$ is tangent to (h), and if Q(v) is a point of inflexion or a cusp of (h), the projection (l) has, at P(v), zero curvature without inflexion (Figures 5.33 and 5.34).

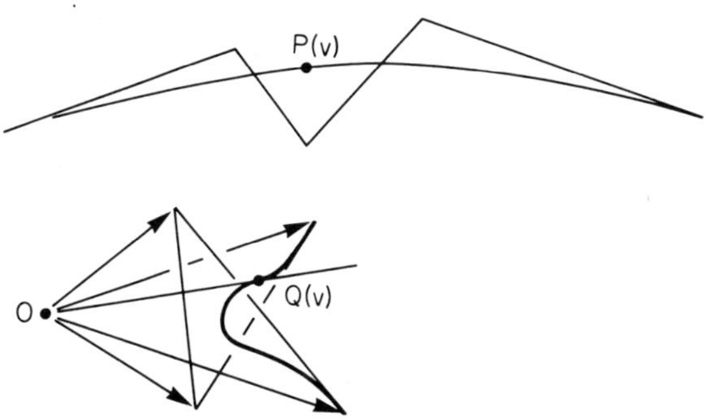

Fig. 5.33

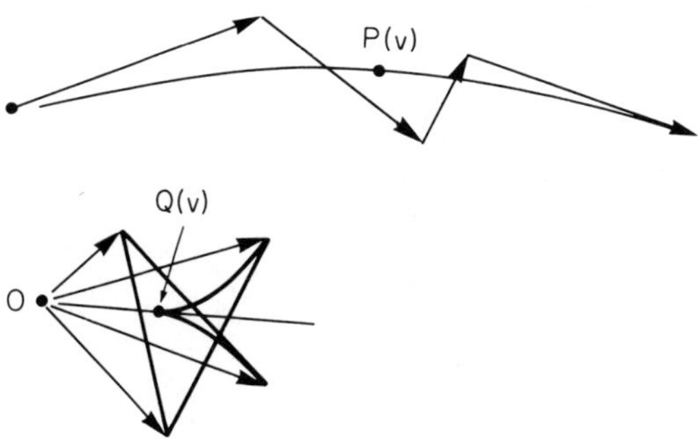

Fig. 5.34

3. If Q(v) is coincident with O, P(v) is a cusp of (l) (Figure 5.35) and the tangents at P(v) and Q(v) are parallel.
4. The characteristic polygon of (h) has one side less than that of (l). If we

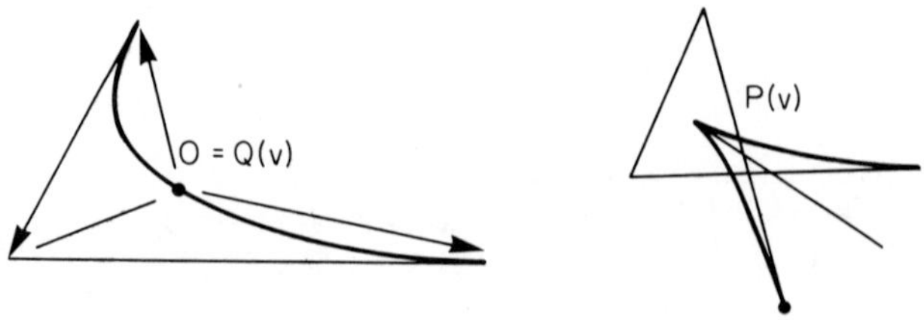

Fig. 5.35

consider the successive hodographs, that of rank s is thus defined by a polygon with $n-s$ sides. The hodographs of rank $n-2$, $n-1$ and n are, respectively, a parabola, a straight line and a point distinct from O. Hodographs of higher rank are points coincident with O.

5. If (l) is defined by a polygon with three vectors, (h) is a parabola.
   (a) If O is outside the parabola, (l) has two points of inflexion, situated relative to P(O) and P(1) as the points of contact are relative to Q(O) and Q(1) (Figure 5.36).

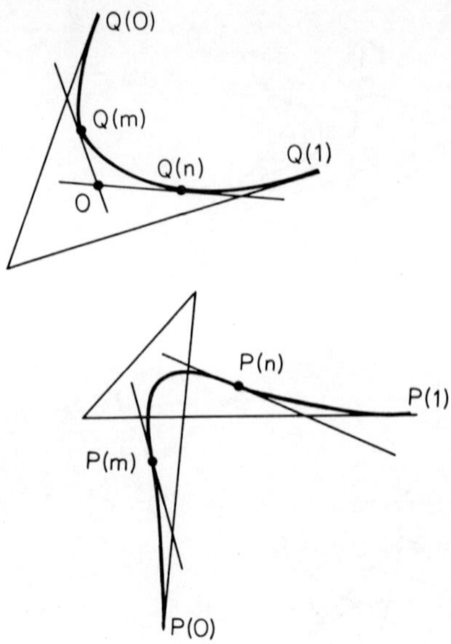

Fig. 5.36

   (b) If O is inside the parabola, (l) has no point of inflexion. If O lies between the parabola and the chord Q(O)Q(1) there are conjugate points on the arc P(O)P(1) where the tangents are parallel and in opposite directions (Figure 5.37).
   (c) At infinity, the tangents to (l) are parallel to the axis of the parabola.

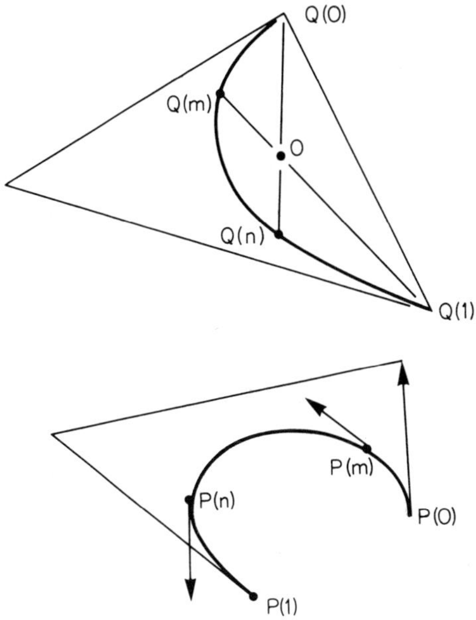

Fig. 5.37

Segmentation of curves

Parametric curves have a particularly interesting property. Given a curve defined by a characteristic polygon whose ends are the points corresponding to the parametric values 0 and 1, if two arbitrary points on this curve are chosen, there exists a characteristic polygon such that, on the same curve, these points correspond respectively to the parametric values 0 and 1, it being understood that the same functions f_i are used in both cases.

Thus given a curve segment bounding several neighbouring surfaces (Figure 5.38), the characteristic polygons of the different sections making up the boundaries may be defined.

In the particular case where the functions defined by equation 5.5 are used, the determination of the new polygons is performed as follows:

For a curve defined by the equation

$$\mathbf{P}(v) = \sum_{0}^{n} \mathbf{a}_i \cdot f_{i}(v) \quad {}_{n}$$

if we consider the points $\mathrm{P}(k_0)$ and $\mathrm{P}(k_1)$, k_0 and k_1 being any values, not necessarily lying between 0 and 1, there is a curve

$$\mathbf{Q}(s) = \sum_{0}^{n} \mathbf{a}'_i \cdot f_{i}(s) \quad {}_{n}$$

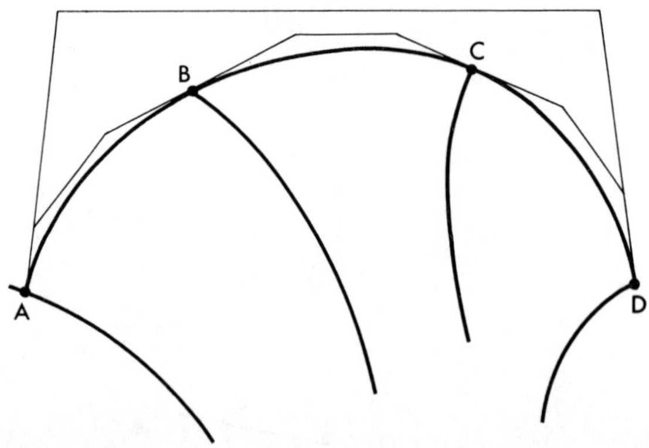

Fig. 5.38

geometrically coincident with the curve $\mathbf{P}(v)$ (Figure 5.39), and such that Q(0) and Q(1) coincide, respectively, with $P(k_0)$ and $P(k_1)$.

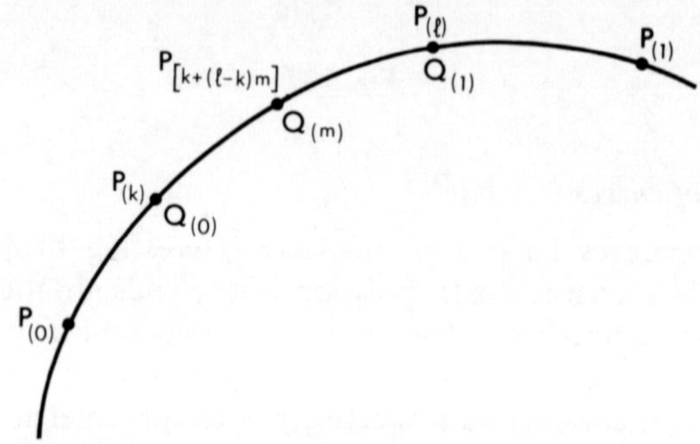

Fig. 5.39

The parameters v and s are related by the linear relation

$$v = k_0 + (k_1 - k_0)s.$$

We have:

$$\mathbf{Q}(s) = \mathbf{P}[k_0 + (k_1 - k_0)s]. \tag{5.27}$$

Let us write $\mathbf{P}(v)$ in the form:

$$\mathbf{P}(v) = \sum_0^n \mathbf{b}_i \,.\, v^i = \sum_0^n \mathbf{b}_i \,.\, [k_0 + (k_1 - k_0)s]^i \tag{5.28}$$

with $\|\mathbf{b}_i\| = \|M_{a,b,n}\| \times \|\mathbf{a}_i\|$, the matrix $\|M_{a,b,n}\|$ being defined by equation 5.10.

Expanding equation 5.28 in the particular case where $n = 3$, we find:

$$\begin{aligned}\mathbf{P}(v) &= \mathbf{b}_0[k_0+(k_1-k_0)s]^0+\mathbf{b}_1[k_0+(k_1-k_0)s]^1 \\ &\quad +\mathbf{b}_2[k_0+(k_1-k_0)s]^2+\mathbf{b}_3[k_0+(k_1-k_0)s]^3 \\ &= (\mathbf{b}_0+\mathbf{b}_1k_0+\mathbf{b}_2k_0^2+\mathbf{b}_3k_0^3)+[\mathbf{b}_1(k_1-k_0)+2\mathbf{b}_2k_0(k_1-k_0) \\ &\quad +3\mathbf{b}_3k_0^2(k_1-k_0)]s+[\mathbf{b}_2(k_1-k_0)^2+3\mathbf{b}_3k_0(k_1-k_0)^2]s^2 \\ &\quad +\mathbf{b}_3(k_1-k_0)^3s^3.\end{aligned}$$

Setting

$$\begin{aligned}\mathbf{b}'_0 &= \mathbf{b}_0+\mathbf{b}_1k_0+\mathbf{b}_2k_0^2+\mathbf{b}_3k_0^3 \\ \mathbf{b}'_1 &= \mathbf{b}_1(k_1-k_0)+2\mathbf{b}_2k_0(k_1-k_0)+3\mathbf{b}_3k_0^2(k_1-k_0) \\ \mathbf{b}'_2 &= \mathbf{b}_2(k_1-k_0)^2+3\mathbf{b}_3k_0(k_1-k_0)^2 \\ \mathbf{b}'_3 &= \mathbf{b}_3(k_1-k_0)^3.\end{aligned}$$

Setting $\|\mathbf{b}'_i\| = \|M_{b,b',n}\| \times \|\mathbf{b}_i\|$ the matrix $\|M_{b,b',n}\|$ has the general form:

$$\|M_{\substack{b,b',n \\ k_0,k_1}}\| = \|k_0^{c-l} \,.\, (k_1-k_0)^{l-1} \,.\, C_{c-1}^{l-1}\|$$

in which l and c define the row and the column containing the general term.

Where $n = 3$, the matrix is

$$\left\|\begin{matrix} 1 & k_0 & k_0^2 & k_0^3 \\ 0 & k_1-k_0 & 2k_0(k_1-k_0) & 3k_0^2(k_1-k_0) \\ 0 & 0 & (k_1-k_0)^2 & 3k_0(k_1-k_0)^2 \\ 0 & 0 & 0 & (k_1-k_0)^3 \end{matrix}\right\|.$$

Knowing the $\mathbf{b}'_i$s, we obtain the $\mathbf{a}'_i$s by means of the inverse matrix of $\|M_{a,b,n}\|$.

Summarizing, the $\mathbf{a}'_i$s (Figure 5.40) are derived from the $\mathbf{a}_i$s by the operation:

$$\|\mathbf{a}'_i\| = \|M_{a,b,n}^{-1}\| \times \|M_{\substack{b,b',n \\ k_0,k_1}}\| \times \|M_{a,b,n}\| \times \|\mathbf{a}_i\|. \qquad (5.29)$$

Modification of the number of sides of a polygon

A curve characterized by a polygon with n sides can also be characterized by a polygon with $n+p$ sides.

In effect, if the curve is defined in the form

$$\mathbf{P}(v) = \sum_0^n \mathbf{b}_i \,.\, v^i$$

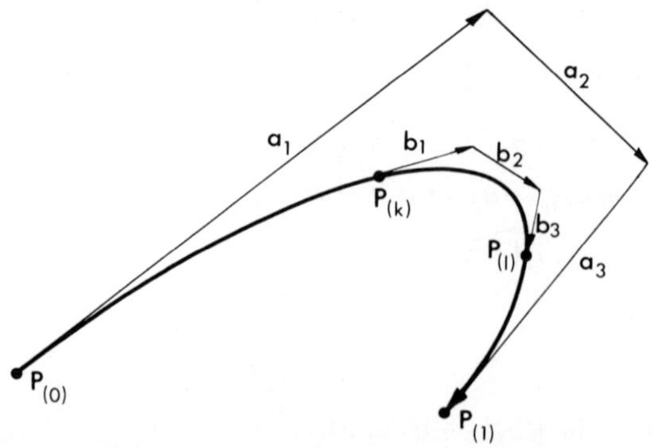

Fig. 5.40

any desired number of zeros may be added to the right of the row matrix of the b_is, obtaining

$$\mathbf{b}_0\mathbf{b}_1 \cdots \mathbf{b}_{n-1}\mathbf{b}_n\underbrace{\mathbf{b}_{n+1} = 0 \cdots \mathbf{b}_{n+p}}_{p \text{ zero terms}} = 0$$

from which we deduce

$$[\mathbf{a}_i'] = [\underset{n+p}{M_{a,b}}]^{-1}[\mathbf{b}_0 \cdots \mathbf{b}_{n+p}].$$

The $\mathbf{a}_i$s make up a polygon with $n+p$ sides.

The reduction in number of the sides of a polygon is only possible under certain conditions.

The vectors $\mathbf{b}_i$ are computed such that

$$\underset{n}{[\mathbf{b}_i]} = \underset{n}{[M_{ab}]}^{-1}\underset{n}{[a_i]}.$$

If the p last vectors are zero or negligible, we can write

$$\sum_0^n \mathbf{b}_i v^i \simeq \sum_0^{n-p} \mathbf{b}_i v^i$$

and the curve defined by the $\underset{n}{\mathbf{a}_i}$s is represented in an approximate way by the $\underset{n-p}{\mathbf{a}_i}$ s.

Descriptive operations

Parametric curves have a fundamental property: their shape depends only on the relative position of the points defining their characteristic vectors, and is independent of the position of the ensemble of the points with respect to the reference system in use.

This property is fundamental, since the reference system for machining is often not identical, nor even parallel, to the reference system of the definition. If the curve is not axis independent, the coordinates of each required point are calculated in the first reference system, and those in the second system are deduced from the former. In the contrary case, it is sufficient to transform the points defining the characteristic vectors from one system to the other, and then to perform all the calculations in the new coordinate system.

Matrix calculations greatly simplify these operations and only some of the fundamental results will be restated here, the reader being referred for the rest to specialized works.

Points. In a two-dimensional coordinate system, a point is represented, for example, by a column matrix composed of its coordinates:

$$[\mathrm{P}] = \begin{bmatrix} x_{\mathrm{P}} \\ y_{\mathrm{P}} \end{bmatrix}$$

Translation. A translation is represented by a column matrix identical to a vector equal to the translation vector and having the same origin as the coordinate system:

$$[\mathrm{T}] = \begin{bmatrix} x_{\mathrm{T}} \\ y_{\mathrm{T}} \end{bmatrix}$$

Performing a translation. The matrix representing a point P_2 after translation is the sum of the matrix representing the point P_1 before translation and the translation matrix:

$$[\mathrm{P}_2] = [\mathrm{P}_1]+[\mathrm{T}].$$

Translation is a commutative operation:

$$[\mathrm{T}_1]+[\mathrm{T}_2] = [\mathrm{T}_2]+[\mathrm{T}_1]$$

Rotation. A rotation about the origin of the two-dimensional coordinate system is represented by a square matrix of the form:

$$[\mathrm{R}_\alpha] = \begin{bmatrix} \cos\alpha & -\sin\alpha \\ \sin\alpha & \cos\alpha \end{bmatrix}$$

Performing a rotation. The matrix representing a point P_2, after the rotation of the point P_1 about the origin of the coordinate system, is the product of the rotation matrix and the point matrix.

$$[\mathrm{P}_2] = [\mathrm{R}_\alpha]\times[\mathrm{P}_1].$$

Rotations about the same axis are commutative.

On the other hand, the order in which a rotation and translation are performed does matter since:

$$[R_\alpha][P_1]+[T] \neq [R_\alpha]\{[P_1]+[T]\}.$$

If a three-dimensional coordinate system is used, the rotation matrix about the z axis is written:

$$\begin{bmatrix} \cos\alpha & -\sin\alpha & 0 \\ \sin\alpha & \cos\alpha & 0 \\ 0 & 0 & 1 \end{bmatrix}.$$

The rotations about the axes OX and OY are represented by matrices with forms analogous to this matrix.

The most general problem consists of performing a rotation about any axis D defined, for example, by the following information:

1. The coordinates (x_A, y_A, z_A) of a point A on the axis D.
2. The angle α between a plane parallel to D and OZ and the plane XOZ.
3. The angle β between D and OZ.

To rotate a point M an angle γ about D, the following operations are performed successively (Figure 5.41):

1. Translation AO: the axis D passes through the origin of the coordinate system.
2. Rotation $-\alpha$ about OZ: the axis D is in the plane XOZ.

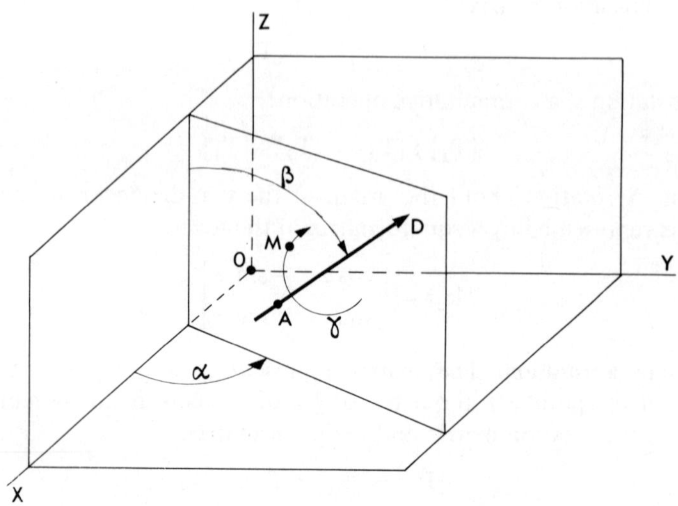

Fig. 5.41

3. Rotation $-\beta$ about OY: the axis D is coincident with OZ.
4. Rotation γ about OZ: the point M performs the prescribed rotation about D.
5. Rotation $+\beta$ about OY:
6. Rotation $+\alpha$ about OZ: the axis D returns to its initial position.
7. Translation $-$AO:

The calculations are represented by the expression:

$$\left\{\begin{bmatrix}\text{Rot}+\alpha\\ \text{OZ}\end{bmatrix}\begin{bmatrix}\text{Rot}+\beta\\ \text{OY}\end{bmatrix}\begin{bmatrix}\text{Rot}+\gamma\\ \text{OZ}\end{bmatrix}\begin{bmatrix}\text{Rot}-\beta\\ \text{OY}\end{bmatrix}\begin{bmatrix}\text{Rot}-\alpha\\ \text{OZ}\end{bmatrix}\right\}$$

$$\times\{[\text{M}]+[\text{T}_{\text{AO}}]\}-[\text{T}_{\text{AO}}].^*$$

Conic projection. For exercises that are based more or less on aesthetic considerations (town planning, car bodies, etc.), it might be desirable to represent the objects in conic perspective and not simply in isometric projection.

Conic perspective may be characterized (Figure 5.42) by the position

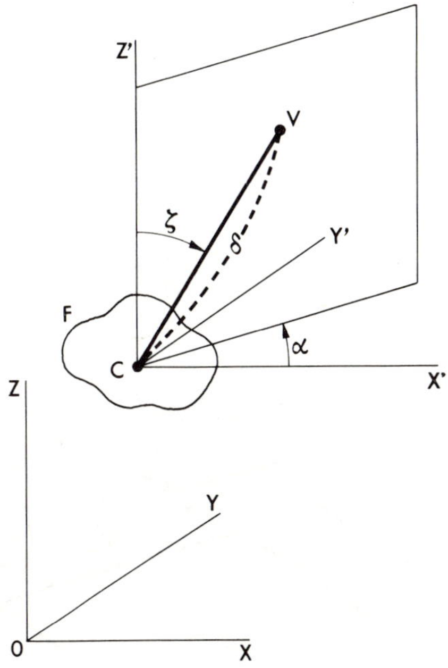

Fig. 5.42

* The resulting matrix is formed by premultiplication of the elementary matrices; this operation is non-commutative.

of the view point V with respect to the centre C of the figure to be projected (F).

This point is defined by the azimuth α, the zenithal distance ξ and the length δ of the line VC.

First, the translation necessary to bring the figure F into a reference system with V for origin and VC for the Z axis is performed, that is to say, we perform in succession:

1. A translation VO.
2. A rotation $(-\alpha, OZ)$.
3. A rotation $(-\xi, OY)$.

For a point P with coordinates:

$$X'_P, Y'_P \quad \text{and} \quad Z'_P$$

in the reference system V, its conic projection, π, on a plane parallel to X'VY' passing through C has coordinates (Figure 5.43)

$$X''_\pi = X'_P \cdot \frac{\delta}{Z'_P}$$

$$Y''_\pi = Y'_P \cdot \frac{\delta}{Z'_P}$$

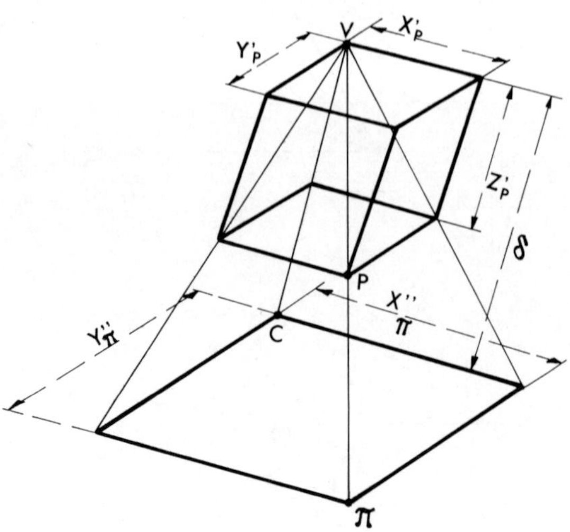

Fig. 5.43

LAWS OF CAMS AND FOLLOWERS

Cams

The existence of numerically controlled camshaft grinders was mentioned on p. 79.

The law of the motion produced by a cam is generally represented by a relation between the displacement of the controlled part, which can be a rotation or a translation, and the angle of rotation of the camshaft:

$$M = f(\alpha). \tag{5.30}$$

In most automobile engines, the cams act on flat tappets; so that if the shape of the cam is defined by equation 5.30, the offset h (Figure 5.44) is:*

$$h = \frac{\Delta m}{\Delta \alpha}.\dagger$$

If the grinding machine (see p. 79) is fitted with a wheelhead mounted on a double slide, the laws of motion are simple to establish, since we have (Figure 5.45)

$$X = n+m+r$$

$$Y = h$$

$$\gamma = \alpha$$

r and n being, respectively, the radii of the grinding wheel and of the base circle of the cam; r occurs only in the adjustment of the origin of X.

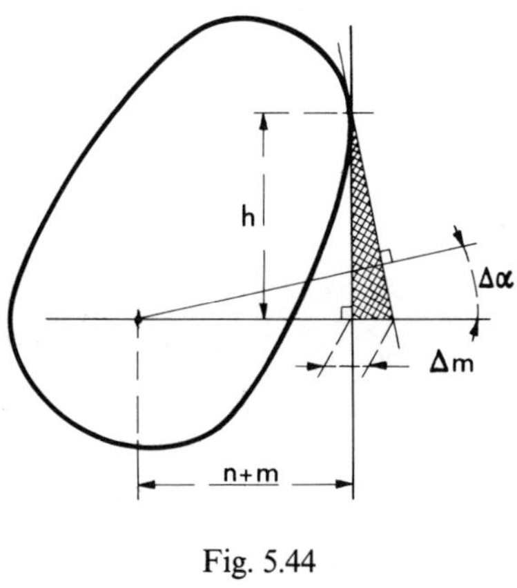

Fig. 5.44

* Note that α is expressed in radians.

† If the law of motion is defined by a table of numbers, there is good reason to use the Δ which designates a finite difference and not the d which denotes differentiation.

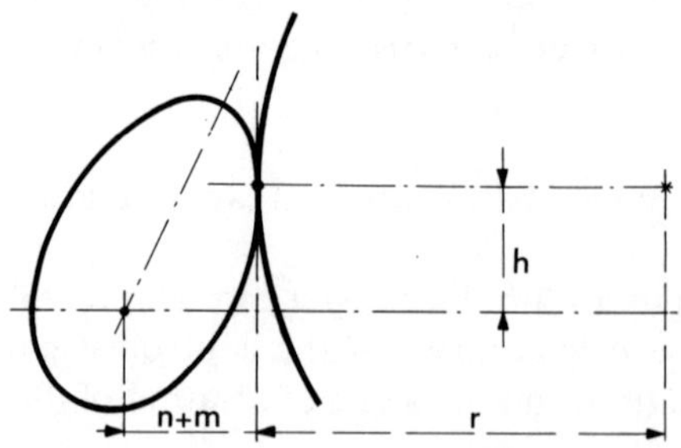

Fig. 5.45

If the grinding machine has only one slide, its displacement takes place in the direction X_1 and the reference system XOY, in which the law of the cam is expressed, oscillates with respect to X_1OY_1, which is the axis of the machine.

The laws of motion are then (Figure 5.46):

$$\rho_1 = \sqrt{(n+m+r)^2+h^2} = OO_1$$

$$\gamma_1 = \gamma+\psi_1$$

with

$$\psi_1 = \tan^{-1}\frac{h}{n+m+r}.$$

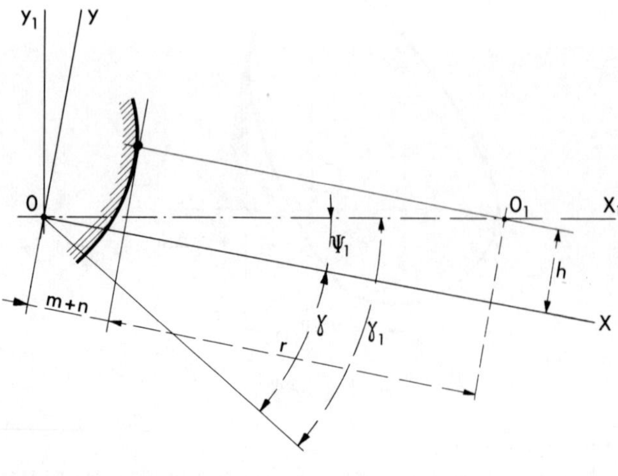

Fig. 5.46

We see that r occurs in these expressions, i.e. the wear of the grinding wheel leads to a variation in the shape of the cam. To overcome this difficulty, several programs corresponding to diverse values of r, chosen as a function of the tolerance on m, are written.

If the flat tappet is replaced by a roller whose centre is displaced on a line perpendicular to the axis of the cam, the law of motion is (Figure 5.47):

$$\mathrm{OA} = f(\gamma) = n+m.$$

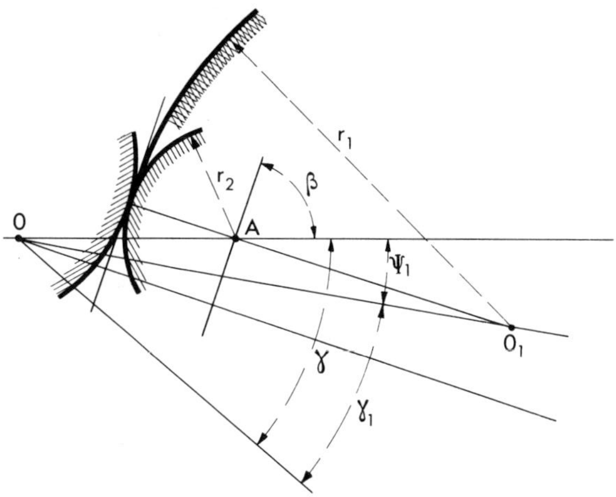

Fig. 5.47

The directrix of the cam is a curve parallel to the trajectory of A in the reference system XOY of the cam.

We have:

$$\tan\beta = (n+m)\frac{\mathrm{d}\gamma}{\mathrm{d}m}.$$

The coordinates of the centre of the roller (O_1) are:

$$\mathrm{OO}_1 = \sqrt{(n+m)^2+(r_1-r_2)^2+2(n+m)(r_1-r_2)\sin\beta}$$

$$\gamma_1 = \gamma-\psi_1$$

$$\psi_1 = \sin^{-1}\left(\frac{r_1-r_2}{\mathrm{OO}_1}\cos\beta\right).$$

The two preceding problems are the most common, but there are other kinematic combinations which can be analysed in a similar way.

Masters

In all large-production-run machines, the part is mounted between centres on an oscillating rocker; the wheel carriage runs on slides perpendicular to the line of centres and serves only to guide the feed of the wheel for the phases of rapid approach, roughing, finishing, 'sparking out' and rapid withdrawal.

Thus the relation between the oscillation of the rocker and the rotation of the part represents the cam law. This movement is controlled by a master cam, integral with the spindle carrying the part, which is, in turn, in contact with a roller fixed with respect to the spindle head (Figure 5.48).

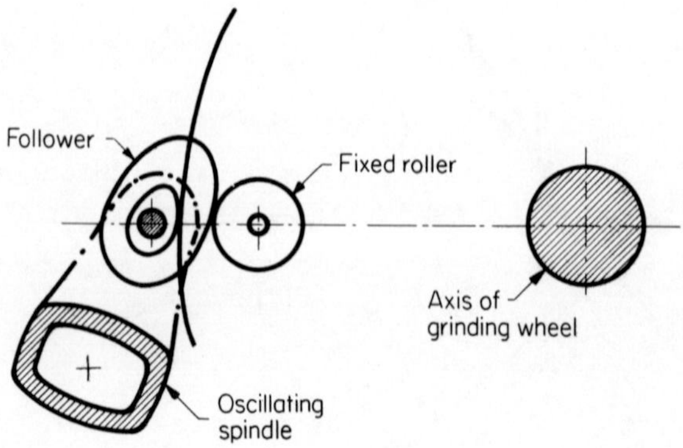

Fig. 5.48

Case 1 : Landis-type machines

The centres of the cam, the roller and the grinding wheel (Figure 5.49) are practically collinear, i.e. the centre of oscillation of the rocker carrying the parts is situated on a perpendicular to OO_1O_2; so that the relative movement of O_1 with respect to O takes place along the axis OX_1.

The movement of O_1 in the coordinate system XOY is defined by the expressions:

$$\left.\begin{aligned} \rho_1 &= \sqrt{(n_1+m+r_1)^2+h^2} \\ \gamma_1 &= \gamma+\psi_1 \\ \psi_1 &= \tan^{-1}\frac{h}{n_1+m+r_1}. \end{aligned}\right\} \qquad (5.31)$$

Setting:

n_2 = radius of the base circle of the master

r_2 = radius of the follower roller

$d = O_2O_1$,

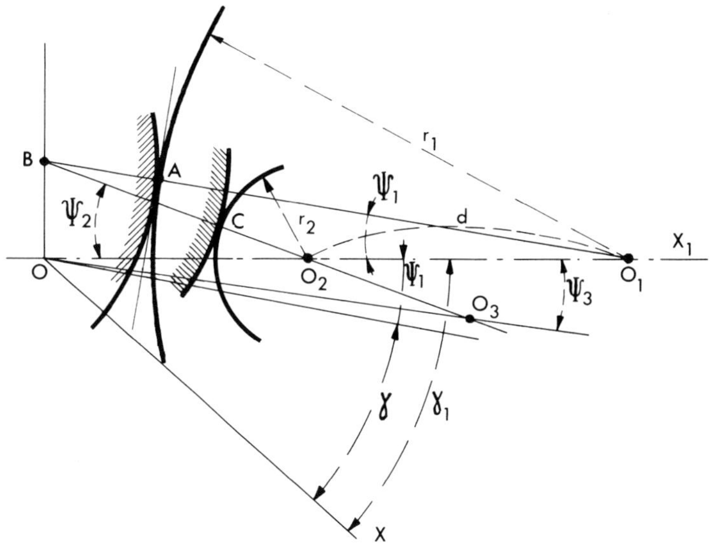

Fig. 5.49

we have:

$$n_1 + r_1 = n_2 + r_2 + d.$$

The law of motion of O_2, the centre of the roller, in the coordinate system XOY is:

$$\rho_2 = OO_2 = \rho_1 - d$$

$$\gamma_2 = \gamma_1 = \gamma + \psi_1.$$

As the master is not produced by a grinding wheel with a radius equal to the radius of the roller, the point of contact of the roller and the master must be found.

The directrix of the master and the locus of O_2 in XOY are two parallel curves separated by a distance r_2; they have the same normal between two corresponding points.

The loci of O_1 and O_2 are two conchoids with parameter d; from a known property, they have the same subnormal at homologous points.

The locus of O_1 and the directrix of the cam are parallel, and their common normal is O_1A. The normal to the locus of O_2 is thus O_2B.

If we then have:

$$\psi_2 = \tan^{-1}\left[\frac{\rho_1}{\rho_2} . \tan \psi_1\right].$$

The point of contact of the roller and the master is thus a point C situated on O_2B such that:

$$O_2C = r_2.$$

On the numerically controlled grinding machine, the master is produced by a grinding wheel of radius r_3, whose centre is situated on CO_2.

The coordinates of O_3 with respect to XOY are:

$$\rho_3 = \overline{OO_3} = \sqrt{\overline{OO_2^2}+\overline{O_2O_3^2}-2\overline{OO_2}\,.\,\overline{OO_3}\cos \widehat{OO_2O_3}}$$

$$\gamma_3 = \gamma+\psi_1-O_2\hat{O}O_3 = \gamma+\psi_1-\psi_3.$$

We have:

$$OO_2 = \rho_2$$

$$O_2O_3 = r_3-r_2$$

$$O\hat{O}_2O_3 = \pi-\psi_2$$

$$\sin\psi_3 = \sin\psi_2\,.\,\frac{O_2O_3}{OO_3}$$

$$= \sin\psi_2\,.\,\frac{r_3-r_2}{\rho_3}.$$

The quantities necessary to calculate ρ_3 and γ_3 are thus available; as this work is accomplished with the aid of a computer, it is carried out step by step, and it would be superfluous to develop ρ_3 and γ_3 as functions of the law $M = f(\alpha)$ and the data n_1, n_2, r_2 and r_3.

Case 2: Norton-type machines

The rocker carrying the parts oscillates about an axis which is not situated on the vertical to O, and the trajectory of this point may be approximated by an oblique line with respect to OO_1, the angle β being about 20°. The movement of the axis of the grinding wheel with respect to the hypothetically fixed axis of the part is thus a line O_1O_1' parallel to the previous line and in the opposite direction (Figure 5.50).

Using the same notation as in the preceding section, we obtain the following results:

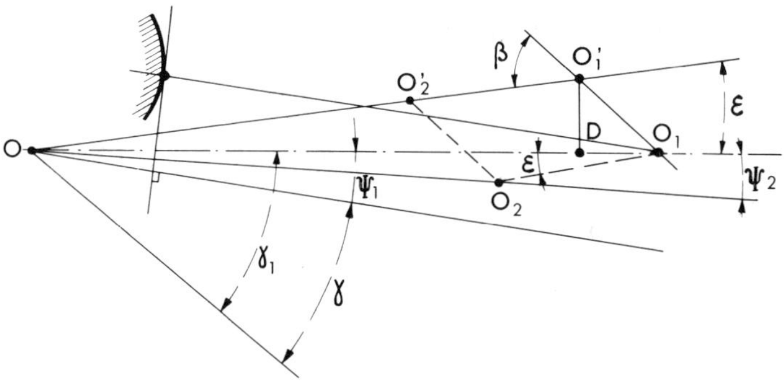

Fig. 5.50

Centre of the grinding wheel

The coordinates of the centre of the grinding wheel O_1 are defined as before, by the expressions:

$$\rho_1 = \sqrt{(n_1+m+r_1)^2+h^2}$$

$$\gamma_1 = \gamma+\psi_1 \qquad (5.32)$$

$$\psi_1 = \tan^{-1}\frac{h}{n_1+m+r_1}.$$

If M denotes the lift which corresponds to the particular position for which O (centre of the cam), O'_1 (centre of the grinding wheel) and O'_2 (centre of the roller) are collinear, and if H is the value of h corresponding to the lift M, we have:

$$OO'_1 = \sqrt{(n_1+M+r_1)^2+H^2} = \rho'.$$

Centre of the roller

When the axis of the grinding wheel is at O'_1, that of the roller is at O'_2, and we have (Figure 5.50):

$$O'_2O'_1 = d$$

$$\varepsilon = \sin^{-1}\frac{O'_1D}{\rho'}$$

ε being small, we have approximately:

$$OD_1 \simeq \rho_1 - \rho'$$

$$O'_1D \simeq (\rho_1 - \rho') . \tan \beta$$

$$\varepsilon = \sin^{-1}\left(\frac{\rho_1 - \rho'}{\rho'} . \tan \beta\right).$$

As the quadrilateral $O_1O'_1O'_2O_2$ is a parallelogram, we have:

$$\sin \psi_2 = \frac{O_1O_2}{OO_2} . \sin \varepsilon.$$

The coordinates of O_2 are thus:

$$\rho_2 = OO_2 = \sqrt{\rho_1^2 + d^2 - 2\rho_1 . d . \cos \varepsilon}$$

$$\gamma_2 = \gamma + \psi - \psi_2.$$

Masters

The tangent of the trajectory of O_2 is given by:

$$\eta = \tan^{-1} \frac{\rho_2 . d\gamma_2}{d\rho_2}$$

or

$$\eta \simeq \tan^{-1} \frac{\rho_2 . \Delta\gamma_2}{\Delta\rho_2}.$$

If the point of contact of the master and the roller is G, we have (Figure 5.51):

$$\gamma_3 = \gamma_2 - \left(\frac{\pi}{2} - \eta\right)$$

$$n_2 + m = OM = OJ - r_2 = OO_2 \sin \eta - r_2$$

$$GH = OO_2 . \cos \eta$$

which defines the law of the master.

As before, it does not seem necessary to develop the formulae.

Twisted curves

MATHEMATICAL CURVES

Mathematical functions are just as suitable for the representation of planar curves (see p. 102) as for twisted curves.

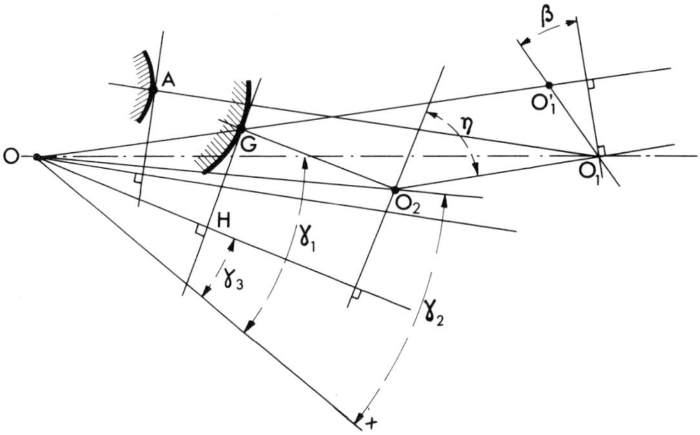

Fig. 5.51

NON-MATHEMATICAL CURVES

Definition of a single curve

The methods described on p. 103 allow the simultaneous definition of two projections of a twisted curve which is the intersection of two cylinders having, for directrices, the two projections.

In this case, the same limitations are found as with plane curves: the impossibility of having a tangent perpendicular to the x axis, and interdependence between the choice of the coordinate system and the shape of the curve.

Piecewise definition

As in the preceding section, a curve may be defined by two of its projections.

However, certain definitions may be impossible: for example (Figure 5.52), given a curve defined by the points (a, a') and (b, b') as well as by the tangents (t, t') and (u, u'), and interpretation of the projections by parabolic arcs gives the solution (c, c'), but the two projections are obviously incompatible.

Parametric curves

The solutions developed on p. 113 are applicable, whatever the number of the coordinates, since the coefficients of the functions are vectors.

The constraints defining an arc may consist of the directions of osculating planes or, equivalently, the binormals, as well as points, tangents and curvatures.

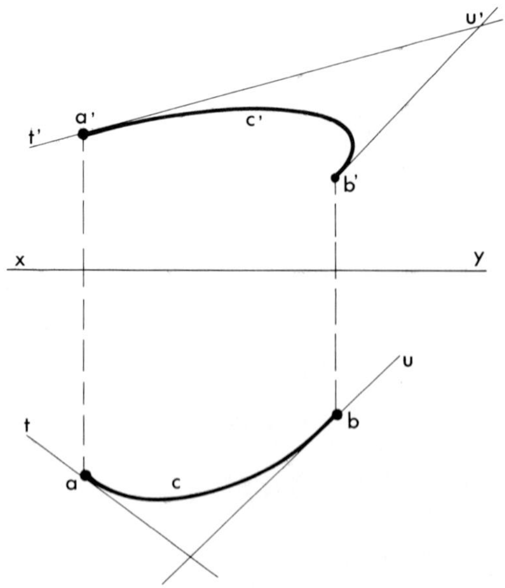

Fig. 5.52

The expression of such a condition being

$$\mathbf{B}_{(v)} = \mathbf{P}'_{(v)} \wedge \mathbf{P}''_{(v)}$$

a bilinear equation is obtained, so that, as for the curvatures (see p. 123), it is generally preferable to increase the number of points and tangents rather than to introduce some conditions represented by equations of degree greater than 1.

The parametric definition of twisted curves, as that of planar curves, is independent of the coordinate system and lends itself readily to descriptive transformations.

Surfaces

MATHEMATICAL SURFACES

The majority of programming languages are capable of describing surfaces such as the plane, the cylinder and the cone of revolution, the sphere and the torus, to which are sometimes added the quadrics, ruled surfaces and surfaces defined biparametrically.[13]

NON-MATHEMATICAL SURFACES

Definition of a single surface

The most conventional method of solving this problem consists of defining (see p. 103) a series of plane sections corresponding to various values of the coordinate perpendicular to the section planes (Figure 5.53).

If we denote by a_{ij} the coefficients of the function:

$$\sum_0^n a_{ij} x^i$$

representing the curve j, there exists, for all values of i, a function of the form:

$$a_i(y) = \sum_0^p b_{ij} \,.\, y^j$$

which takes the value a_{ij} when y is the value of the plane coordinate corresponding to the curve j.

Thus, for each pair of coordinates x and y, there is a corresponding value of z:

$$z(x, y) = \sum_{i=0}^{n} \left[\sum_{j=0}^{p} b_{ij} \,.\, y^j \right] . x^i .$$

The difficulties mentioned on p. 103 with regard to curves occur again with surfaces. It is impossible to have a tangent parallel to OZ, and the shape of the surface is related to the choice of coordinate axes, demanding lengthy calculations if a rotation is to be performed.

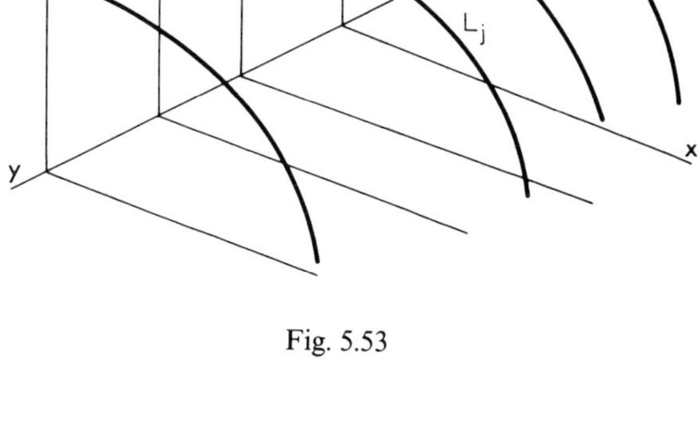

Fig. 5.53

There are, however, areas where this method is suitable and mention must be made of M. Fayard's use of it for defining, in particular, the shapes of turbine blades.[18,19] Instead of using polynomials, he has used harmonic functions, and the surface is represented by an equation of the form:

$$z(x, y) = \sum_{i=0}^{n} \left\{ \left[\sum_{j=0}^{p} (a_{ij} \sin j\psi + b_{ij} \cos j\psi) \right] \sin i\phi + \left[\sum_{j=0}^{p} (c_{ij} \sin j\psi + d_{ij} \cos j\psi) \right] \cos i\phi \right\}.$$

The limits n and p are fixed in accordance with the tolerance on the shape of the profiles. For current problems, their value does not exceed 8.

Piecewise definition

The Inaba method

Dr. Inaba, Director at the Fujitsu Ltd., has perfected a method* based on the following principle:

A surface (Figure 5.54) is divided into squares or 'patches' bounded by planes parallel to two faces of the coordinate system, for example XOZ and YOZ.

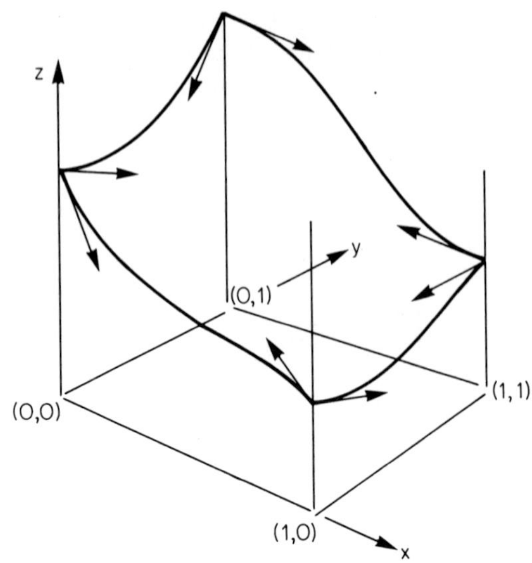

Fig. 5.54

* Designated by the name 'F Mesh'.

By translation and scaling, each patch is represented in a system such that the coordinates x and y lie between 0 and 1:

$$0 \leq \left| \begin{matrix} x \\ y \end{matrix} \right| \leq 1.$$

Under this assumption, the surface is represented by the relation

$$z_{xy} = \sum_{0,0}^{3,3} a_{ij} x^i y^j \qquad (0 \leq i, j \leq 3).$$

Each section of this surface by a plane parallel to XOZ or YOZ is a cubic parabola with axis parallel to OZ.

Isolated patch

To determine the sixteen coefficients $a_{00}, \ldots, a_{33}$, some constraints derived from the z coordinates and the direction of the normal at each corner of the square are imposed.

In practice, it is relatively easy to measure the coordinate of a point situated on an experimental surface, but much more difficult to determine the direction of the normal; thus it is preferable to define at these points the tangents to the curves of intersection of the planes parallel to those of the coordinate axes. Use is then made of the procedures described on p. 105.

The coefficients a_{ij} must thus satisfy first the following twelve conditions:

$$z(0, 0) = a_{00} \tag{5.33}$$

$$\frac{\partial z}{\partial x}(0, 0) = a_{10} \tag{5.34}$$

$$\frac{\partial z}{\partial y}(0, 0) = a_{01} \tag{5.35}$$

$$z(1, 0) = a_{00} + a_{10} + a_{20} + a_{30} \tag{5.36}$$

$$\frac{\partial z}{\partial x}(1, 0) = a_{10} + 2a_{20} + 3a_{30} \tag{5.37}$$

$$\frac{\partial z}{\partial y}(1, 0) = a_{01} + a_{11} + a_{21} + a_{31} \tag{5.38}$$

$$z(0, 1) = a_{00} + a_{01} + a_{02} + a_{03} \tag{5.39}$$

$$\frac{\partial z}{\partial x}(0, 1) = a_{10} + a_{11} + a_{12} + a_{13} \tag{5.40}$$

$$\frac{\partial z}{\partial y}(0, 1) = a_{01} + 2a_{02} + 3a_{03} \tag{5.41}$$

$$z(1,1) = a_{00} + \cdots + a_{33} \tag{5.42}$$

$$\begin{aligned}\frac{\partial z}{\partial x}(1,1) = {} & (a_{10}+a_{11}+a_{12}+a_{13}) \\ & +2(a_{20}+a_{21}+a_{22}+a_{23}) \\ & +3(a_{30}+a_{31}+a_{32}+a_{33})\end{aligned} \tag{5.43}$$

$$\begin{aligned}\frac{\partial z}{\partial y}(1,1) = {} & (a_{01}+a_{11}+a_{21}+a_{31}) \\ & +2(a_{02}+a_{12}+a_{22}+a_{32}) \\ & +3(a_{03}+a_{13}+a_{23}+a_{33}).\end{aligned} \tag{5.44}$$

The system of twelve equations must be completed by four arbitrary conditions.

For example, the surface may be made to pass through four arbitrary points, provided that these are not situated on the boundary curves, i.e. their x and y coordinates should be neither 0 nor 1.

Also, an arbitrary value may be given to four coefficients, provided that they are not:

a_{00}, a_{10} and a_{01} (defined, respectively, by equations 5.33, 5.34 and 5.35)
a_{20} and a_{30} (roots of the system formed by equations 5.36 and 5.37)
a_{02} and a_{03} (roots of the system formed by equations 5.39 and 5.41).

Patch tangent to a given patch

Consider a surface S′ (Figure 5.55) tangent along the line CD to a surface S which has been previously defined.

For the line CD to be common to S and S′, we must have:

$$z'(0,0) = z(1,0)$$

$$\frac{\partial z'}{\partial y}(0,0) = \frac{\partial z}{\partial y}(1,0)$$

$$z'(0,1) = z(1,1)$$

$$\frac{\partial z'}{\partial y}(0,1) = \frac{\partial z}{\partial y}(1,1).$$

Furthermore, for the two surfaces to be tangents along CD, we must have:

$$\frac{\partial z'}{\partial x}(0,y) \equiv \frac{\partial z}{\partial x}(1,y) \tag{5.45}$$

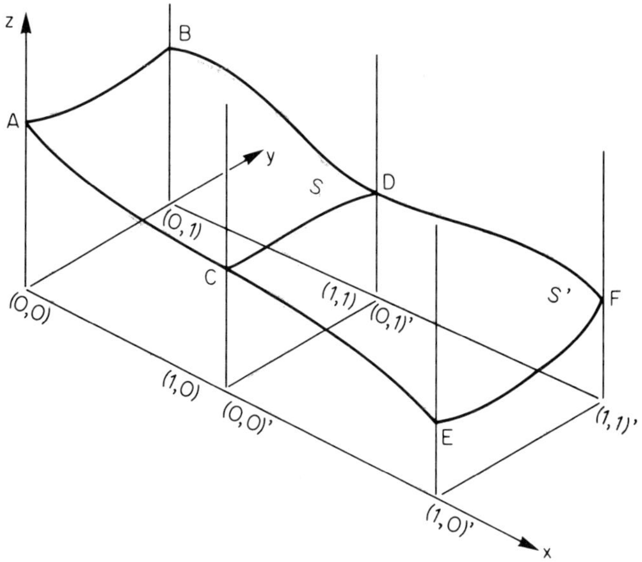

Fig. 5.55

or

$$a'_{10}+a'_{11}y+a'_{12}y^2+a'_{13}y^3$$
$$\equiv (a_{10}+2a_{20}+3a_{30})+(a_{11}+2a_{21}+3a_{31})y$$
$$+(a_{12}+2a_{22}+3a_{32})y^2+(a_{13}+2a_{23}+3a_{33})y^3.$$

For the two polynomials to be identical, it is necessary and sufficient for their coefficients to be identical, giving the conditions:

$$a'_{10} = a_{10}+2a_{20}+3a_{30} \quad (5.46)$$

$$a'_{11} = a_{11}+2a_{21}+3a_{31} \quad (5.47)$$

$$a'_{12} = a_{12}+2a_{22}+3a_{32} \quad (5.48)$$

$$a'_{13} = a_{13}+2a_{23}+3a_{33}. \quad (5.49)$$

The conditions of continuity of AC and BD with CE and DF, i.e.

$$\left.\begin{aligned}\frac{\partial z'}{\partial x}(0,0) &= \frac{\partial z}{\partial x}(1,0)\\ \frac{\partial z'}{\partial x}(0,1) &= \frac{\partial z}{\partial x}(1,1)\end{aligned}\right\} \quad (5.50)$$

are naturally included in equations 5.46 to 5.49, since they represent, in equation 5.45 the particular case where y is equal to 0 or 1.*

The patch S′ is thus subjected to fourteen conditions:

(a) Six conditions relating to the points E(1, 0) and F(1,1):

$$z'(1,0) = a'_{00}+a'_{10}+a'_{20}+a'_{30} \tag{5.51}$$

$$\frac{\partial z'}{\partial x}(1,0) = a'_{10}+2a'_{20}+3a'_{30} \tag{5.52}$$

$$\frac{\partial z'}{\partial y}(1,0) = a'_{01}+a'_{11}+a'_{21}+a'_{31} \tag{5.53}$$

$$z'(1,1) = a'_{00}+\cdots+a'_{33} \tag{5.54}$$

$$\begin{aligned}\frac{\partial z'}{\partial x}(1,1) = {} & (a'_{10}+a'_{11}+a'_{12}+a'_{13}) \\ & +2(a'_{20}+a'_{21}+a'_{22}+a'_{23}) \\ & +3(a'_{30}+a'_{31}+a'_{32}+a'_{33})\end{aligned} \tag{5.55}$$

$$\begin{aligned}\frac{\partial z'}{\partial y}(1,1) = {} & (a'_{01}+a'_{11}+a'_{21}+a'_{31})+2(a'_{02}+a'_{12}+a'_{22}+a'_{32}) \\ & +3(a'_{03}+a'_{13}+a'_{23}+a'_{33}).\end{aligned} \tag{5.56}$$

(b) Four conditions relating to the line CD:

$$z'(0,0) = z(1,0)$$

$$\frac{\partial z'}{\partial y}(0,0) = \frac{\partial z}{\partial y}(1,0)$$

$$z'(0,1) = z(1,1)$$

$$\frac{\partial z'}{\partial y}(0,1) = \frac{\partial z}{\partial y}(1,1).$$

These four conditions are, respectively, equivalent to:

$$a'_{00} = a_{00}+a_{10}+a_{20}+a_{30} \tag{5.57}$$

$$a'_{01} = a_{01}+a_{11}+a_{21}+a_{31} \tag{5.58}$$

$$a'_{00}+a'_{01}+a'_{02}+a'_{03} = a_{00}+\cdots+a_{33} \tag{5.59}$$

$$\begin{aligned}a'_{01}+2a'_{02}+3a'_{03} = {} & (a_{01}+a_{11}+a_{21}+a_{31}) \\ & +2(a_{02}+a_{12}+a_{22}+a_{32}) \\ & +3(a_{03}+a_{13}+a_{23}+a_{33}).\end{aligned} \tag{5.60}$$

* Equation 5.50 is obtained by adding equations 5.46, 5.47, 5.48 and 5.49 term by term.

(c) Four conditions (5.46 to 5.49) relating to continuity along the line CD.

Altogether, fourteen conditions are imposed, and, in addition, two arbitrary conditions may be fixed, represented, for example, by two points through which the surface must pass.

Patch lying between two given patches

First case

The patches S and S″ (Figure 5.56) being predefined, the patch S′ is subjected to eight conditions for continuity with S along CD and also for continuity with S″ along EF.

These sixteen coefficients may thus be determined without using additional arbitrarily chosen conditions.

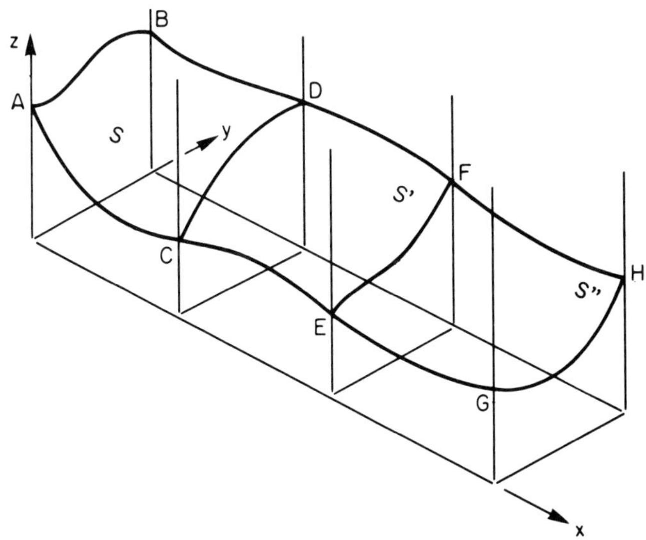

Fig. 5.56

Second case

The patches S and S″ (Figure 5.57) being predefined, the patch S′ should be continuous with S along DE and with S″ along DG.

The patch S′ is thus subjected to the following conditions:

(a) Three conditions relating to the point H(1, 1). These are represented by equations 5.54, 5.55 and 5.56.
(b) Four conditions relating to the line DE. These are equations 5.57, 5.58, 5.59 and 5.60.

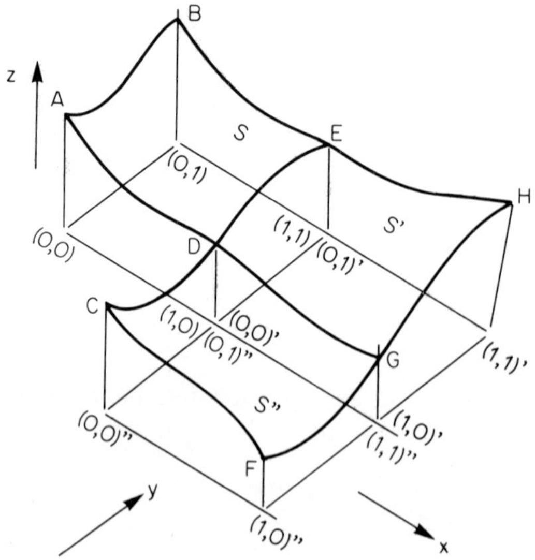

Fig. 5.57

(c) Four conditions relating to continuity along DE. These are expressed by the equations 5.46, 5.47, 5.48 and 5.49.

(d) Four conditions relating to the line DG. These are:

$$z'(0, 0) = z''(0, 1) \tag{5.61}$$

$$\frac{\partial z'}{\partial x}(0, 0) = \frac{\partial z''}{\partial x}(0, 1) \tag{5.62}$$

$$z'(1, 0) = z''(1, 1) \tag{5.63}$$

$$\frac{\partial z'}{\partial x}(1, 0) = \frac{\partial z''}{\partial x}(1, 1). \tag{5.64}$$

As we have, of necessity,

$$z''(0, 1) = z(1, 0)$$

and

$$\frac{\partial z''}{\partial x}(0, 1) = \frac{\partial z}{\partial x}(1, 0)$$

condition 5.61 is equivalent to equation 5.57.

Condition 5.62 is satisfied if the conditions 5.46, 5.47, 5.48 and 5.49 are also satisfied.

The conditions 5.63 and 5.64 are, respectively, equivalent to:

$$a'_{00}+a'_{10}+a'_{20}+a'_{30} = a''_{00}+\cdots+a''_{33} \tag{5.65}$$

$$\begin{aligned} a'_{10}+2a'_{20}+3a'_{30} = {} & (a''_{10}+a''_{11}+a''_{12}+a''_{13}) \\ & +2(a''_{20}+a''_{21}+a''_{22}+a''_{23}) \\ & +3(a''_{30}+a''_{31}+a''_{32}+a''_{33}). \end{aligned} \tag{5.66}$$

(e) Four conditions relating to continuity along DG. The general condition is

$$\frac{\partial z'}{\partial y}(x,0) \equiv \frac{\partial z''}{\partial y}(x,1) \tag{5.67}$$

or

$$\begin{aligned} a'_{01}+a'_{11}x+a'_{21}x^2+a'_{31}x^3 \equiv {} & (a''_{01}+2a''_{02}+3a''_{03})+(a''_{11}+2a''_{12}+3a''_{13})x \\ & +(a''_{21}+2a''_{22}+3a''_{23})x^2 \\ & +(a''_{31}+2a''_{32}+3a''_{33})x^3. \end{aligned} \tag{5.68}$$

By equating the coefficients of the two polynomials in x we obtain the four conditions:

$$a'_{01} = a''_{01}+2a''_{02}+3a''_{03} \tag{5.69}$$

$$a'_{11} = a''_{11}+2a''_{12}+3a''_{13} \tag{5.70}$$

$$a'_{21} = a''_{21}+2a''_{22}+3a''_{23} \tag{5.71}$$

$$a'_{31} = a''_{31}+2a''_{32}+3a''_{33}. \tag{5.72}$$

Equations 5.58 and 5.69 are, respectively, equivalent to

$$\frac{\partial z'}{\partial y}(0,0) = \frac{\partial z}{\partial y}(1,0)$$

$$\frac{\partial z'}{\partial y}(0,0) = \frac{\partial z''}{\partial y}(0,1).$$

As the surfaces S and S″ have the same tangent plane at D, we deduce that

$$\frac{\partial z}{\partial y}(1,0) \equiv \frac{\partial z''}{\partial y}(0,1).$$

Thus, if equation 5.58 is satisfied, then so also is equation 5.69.

Moreover, we confirm that, if the surfaces S and S″ have the same tangent plane at D, we have:

$$a_{11}+2a_{21}+3a_{31} = a''_{11}+2a''_{12}+3a''_{13}$$

and as a result equation 5.70 is satisfied when equation 5.47 is also satisfied.

Summing up, the conditions imposed on the surface S′ are reduced to:

equations 5.54, 5.55 and 5.56 for the point H,
equations 5.57, 5.58, 5.59 and 5.60 for the line DE,
equations 5.46, 5.47, 5.48 and 5.49 for continuity along DE,
equations 5.63, and 5.64 for the line DG,
equations 5.71 and 5.72 for continuity along DG,

making a total of fifteen conditions.

Thus one arbitrary condition may be imposed on the surface S′, for example, that it should pass through an arbitrary point which does not lie on the boundary curves.

Thus, among the $m \times n$ patches making up a surface, there are (Figure 5.58):

one on which four arbitrary conditions may be imposed,
$(m+n-2)$ on which two arbitrary conditions may be imposed and
$(m-1)(n-1)$ on which one arbitrary condition may be imposed,

making a total of $(m+1)(n+1)$ arbitrary constraints.

The inventors of the method believe that, in the majority of cases, only twelve terms need be used, the discontinuity between the surfaces being imperceptible.

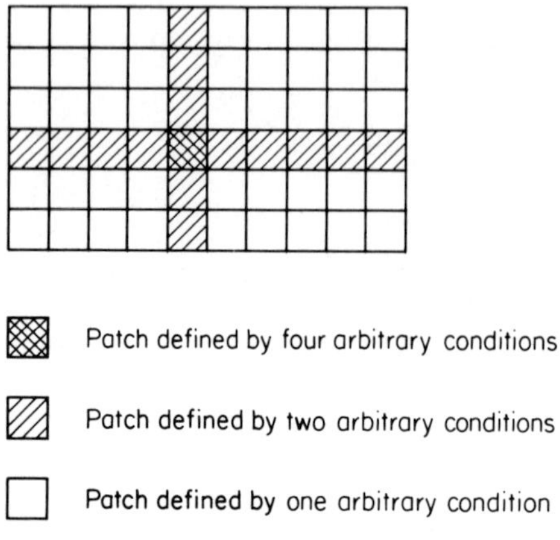

Fig. 5.58

This method has been made the basis of a very complete system, called F Mesh, by Fujitsu, in which a great number of special cases have been resolved. We have been able here only to give a brief description limited to the fundamental principle.

The calculations were first programmed for the FANUC computer, made by Fujitsu, but the program is available now on other computers.

Definition by interpolating functions

S. A. Coons's method

Professor Steven Anson Coons* has devised a mathematical method which ensures continuity between neighbouring patches.[20]

First-order continuity. Four curves bounding an elementary surface are defined (Figure 5.59) by the expressions:

$$H_0(v),\ H_1(v),\ J_0(u) \text{ and } J_1(u).$$

Between their points of intersection, the parameters u and v vary between the values 0 and 1. These four points are expressed respectively:

$$H_0(0) \quad \text{or} \quad J_0(0)$$
$$H_0(1) \quad \text{or} \quad J_1(0)$$
$$H_1(0) \quad \text{or} \quad J_0(1)$$
$$H_1(1) \quad \text{or} \quad J_1(1).$$

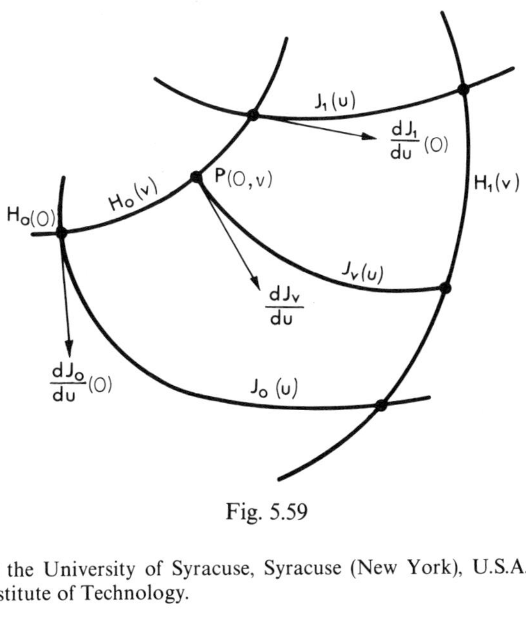

Fig. 5.59

* Professor at the University of Syracuse, Syracuse (New York), U.S.A., formerly at the Massachusetts Institute of Technology.

Let F_0 and F_1 be two 'blending functions'

$$F_0 = 2t^3 - 3t^2 + 1$$

$$F_1 = -2t^3 + 3t^2$$

where t is some variable.

A point on this surface is defined by u and v

$$(0 \leq u \leq 1, 0 \leq v \leq 1)$$

according to the equation:

$$\mathbf{P}(u, v) = \mathbf{H}_0(v) \,.\, F_0(u) + \mathbf{H}_1(v) \,.\, F_1(u) + \mathbf{J}_0(u) \,.\, F_0(v) + \mathbf{J}_1(u) \,.\, F_1(v)$$

$$- \mathbf{H}_0(0) \,.\, F_0(u) F_0(v) - \mathbf{H}_0(1) \,.\, F_0(u) \,.\, F_1(v)$$

$$- \mathbf{H}_1(0) F_1(u) F_0(v) - \mathbf{H}_1(1) F_1(u) F_1(v).$$

A point P(0, v) has coordinates 0 and v ($0 \leq v \leq 1$), and, since we have:

$$F_0(0) = 1$$

$$F_1(0) = 0$$

as a result:

$$\mathbf{P}(0, v) = \mathbf{H}(0, v)$$

and this point is indeed situated on the line $\mathbf{H}_0(v)$.

Through this point passes a generatrix $J_v(u)$ of the surface, defined by v = constant.

At $\mathbf{P}(0, v)$, the tangent to this generatrix is expressed by:

$$\mathbf{P}'(0, v) = \frac{\partial \mathbf{P}}{\partial u}(0, v).$$

We have:

$$\frac{\partial \mathbf{P}}{\partial u}(u, v) = \mathbf{H}_0(v) \,.\, \frac{dF_0(u)}{du} + \mathbf{H}_1(v) \,.\, \frac{dF_1(u)}{du}$$

$$+ \frac{d\mathbf{J}_0}{du}(u) \,.\, F_0(v) + \frac{d\mathbf{J}_1}{du}(u) \,.\, F_1(v)$$

$$- \mathbf{H}_0(0) \,.\, \frac{dF_0}{du}(u) F_0(v) - \mathbf{H}_0(1) \,.\, \frac{dF_0}{du}(u) \,.\, F_1(v)$$

$$- \mathbf{H}_1(0) \,.\, \frac{dF_1}{du}(u) \,.\, F_0(v) - \mathbf{H}_1(1) \,.\, \frac{dF_1}{du}(u) \,.\, F_1(v).$$

In the particular case of P(0, v):

$$\frac{\partial \mathbf{P}}{du}(0, v) = \mathbf{H}_0(v) \,.\, \frac{\mathrm{dF}_0}{\mathrm{d}u}(0) + \mathbf{H}_1(v)\frac{\mathrm{dF}_1}{\mathrm{d}u}(0) + \frac{\mathrm{d}\mathbf{J}_0}{\mathrm{d}u}(0) \,.\, \mathrm{F}_0(v) + \frac{\mathrm{d}\mathbf{J}_1}{\mathrm{d}u}(0) \,.\, \mathrm{F}_1(v)$$

$$-\mathbf{H}_0(0) \,.\, \frac{\mathrm{dF}_0}{\mathrm{d}u}(0) \,.\, \mathrm{F}_0(v) - \mathbf{H}_0(1) \,.\, \frac{\mathrm{dF}_0}{\mathrm{d}u}(0) \,.\, \mathrm{F}_1(v)$$

$$-\mathbf{H}_1(0) \,.\, \frac{\mathrm{dF}_1}{\mathrm{d}u}(0) \,.\, \mathrm{F}_0(v) - \mathbf{H}_1(1) \,.\, \frac{\mathrm{dF}_1}{\mathrm{d}u}(0) \,.\, \mathrm{F}_1(v).$$

As we have:

$$\frac{\mathrm{dF}_0}{\mathrm{d}u}(0) \equiv \frac{\mathrm{dF}_0}{\mathrm{d}u}(1) \equiv \frac{\mathrm{dF}_1}{\mathrm{d}u}(0) \equiv \frac{\mathrm{dF}_1}{\mathrm{d}u}(1) \equiv 0$$

therefore:

$$\frac{\partial \mathbf{P}}{\partial u}(0, v) = \frac{\mathrm{d}\mathbf{J}_0}{\mathrm{d}u}(0) \,.\, \mathrm{F}_0(v) + \frac{\mathrm{d}\mathbf{J}_1}{\mathrm{d}u}(0) \,.\, \mathrm{F}_1(v).$$

Now: $\mathrm{d}\mathbf{J}_0/\mathrm{d}u(0)$ and $\mathrm{d}\mathbf{J}_1/\mathrm{d}u(0)$ are, respectively, the tangent vectors at $\mathrm{H}_0(0)$ and $\mathrm{H}_0(1)$ to the curves $\mathrm{J}_0(u)$ and $\mathrm{J}_1(u)$.

Thus the tangent vector to the curve $\mathrm{J}_v(u)$ at the point $\mathrm{P}_0(v)$ depends only on the tangent vectors to $\mathrm{J}_0(u)$ and $\mathrm{J}_1(u)$ at $\mathrm{H}_0(0)$ and $\mathrm{H}_0(1)$.

The same proof applies to the three other sides of S.

Let T be a second surface defined as for S by four curves (Figure 5.60):

$$\mathrm{K}_0(v),\ \mathrm{K}_1(v),\ \mathrm{L}_0(u) \quad \text{and} \quad \mathrm{L}_1(u).$$

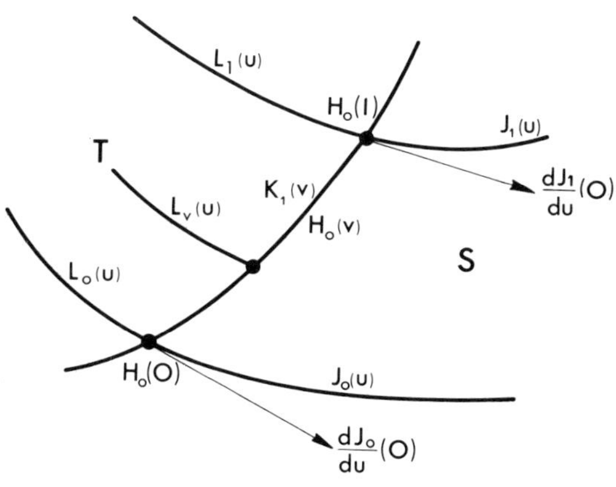

Fig. 5.60

If the curves $L_0(u)$ and $J_0(u)$ are tangential at $H_0(0)$, we have:

$$\frac{d\mathbf{J}_0}{du}(0) = m_0 \frac{d\mathbf{L}_0}{du}(1)$$

similarly at $H_0(1)$, we have

$$\frac{d\mathbf{J}_1}{du}(0) = m_1 \frac{d\mathbf{L}_1}{du}(1)$$

m_0 and m_1 being positive coefficients.

At a point $K_1(v)$, coincident with the point $H_0(v)$, the tangent to the curve $L_v(u)$ is given by:

$$\frac{d\mathbf{L}_v}{du}(u) = \frac{d\mathbf{L}_0}{du}(1) \,.\, F_0(v) + \frac{d\mathbf{L}_1}{du}(1) \,.\, F_1(v).$$

For the two surfaces S and T to be tangent at $H_0(v)$, it is sufficient that:

$$\frac{d\mathbf{L}_v}{du}(u) \quad \text{and} \quad \frac{d\mathbf{J}_v}{du}(u)$$

should be collinear.

Now:

$$\frac{d\mathbf{L}_v}{du}(u) = \frac{1}{m_0} \cdot \frac{d\mathbf{J}_0}{du}(0) \,.\, F_0(v) + \frac{1}{m_1} \cdot \frac{d\mathbf{J}_1}{du}(0) F_1(v)$$

For this condition to be satisfied, it suffices that $m_0 = m_1$, i.e. the vectors

$$\frac{d\mathbf{L}_0}{du}(1) \quad \text{and} \quad \frac{d\mathbf{L}_1}{du}(1)$$

should be proportional, respectively, to

$$\frac{d\mathbf{J}_1}{du}(0) \quad \text{and} \quad \frac{d\mathbf{J}_1}{du}(0).$$

This condition is sufficient, but not necessary. In effect, for S and T to be tangential at $H_0(v)$, it would be sufficient for the vectors:

$$\frac{d\mathbf{J}_v}{du}(u), \frac{d\mathbf{L}_v}{du}(u) \quad \text{and} \quad \frac{d\mathbf{H}_0}{dv}(v)$$

to be coplanar, or, in other words, for their triple scalar product to be zero.

The latter condition leads to the setting up of an equation whose solution is extremely laborious. For this reason, the proportionality, or even the identity of the derivative vectors at the end points of the lines making up the mesh, are generally considered adequate.

Second-order continuity. The functions F_0 and F_1 mentioned earlier ensure that the adjacent surfaces are tangent along their common boundaries.

If it is necessary, for example, for reasons related to fluid dynamics, to ensure second-order continuity, we replace F_0 and F_1 by G_0 and G_1, respectively:

$$G_0 = -6t^5+15t^4-10t^3+1$$
$$G_1 = 6t^5-15t^4+10t^3.$$

We see that, as for F_0 and F_1, we have:

$$G_0'(0) \equiv G_0'(1) \equiv G_1'(0) \equiv G_1'(1) \equiv 0$$

and furthermore:

$$G_0''(0) \equiv G_0''(1) \equiv G_1''(0) \equiv G_1''(1) \equiv 0.$$

With these conditions, if the curves $L_0(u)$ and $L_1(u)$ are, respectively, osculatory to the curves $J_0(u)$ and $J_1(u)$ at the points $H_0(0)$ and $H_1(1)$, each curve $L_v(u)$ is osculatory to the curve $J_v(u)$ at the point $H_0(v)$.

These conditions, for the point $H_0(0)$, are expressed:

$$\mathbf{J}_0(0) = \mathbf{L}_0(1) \qquad \text{common end point}$$

$$\frac{d\mathbf{J}_0}{du}(0) = m \cdot \frac{d\mathbf{L}_0}{du}(1) \qquad \text{common tangent, } m \text{ being a positive coefficient}$$

$$\frac{d\mathbf{J}_0}{du}(0) \wedge \frac{d^2\mathbf{J}_0}{du^2}(0) = m'\frac{d\mathbf{L}_0}{du}(1) \wedge \frac{d^2\mathbf{L}_0}{du^2}(1) \qquad \text{common binormal, } m' \text{ being a positive coefficient}$$

$$\frac{|d\mathbf{J}_0(0)|^3}{\left|\frac{d\mathbf{J}_0}{du}(0) \wedge \frac{d^2\mathbf{J}_0}{du^2}(0)\right|} = \frac{|d\mathbf{L}_0(1)|^3}{\left|\frac{d\mathbf{L}_0}{du}(1) \wedge \frac{d^2\mathbf{L}_0}{du^2}(1)\right|} \qquad \text{equality of radii of curvature.}$$

Analogous relations apply to the curves $J_1(u)$ and $L_1(u)$ at the point $H_0(1)$.

By developing these calculations, we obtain similar relations for the curves

$$J_v(u) \quad \text{and} \quad L_v(u) \quad \text{at} \quad H_0(v)$$

It may thus be seen that Professor Coons's method has the great advantage of ensuring first- or second-order continuity, whatever modifications are made to the curves of the grid which defines the overall surface.

On the other hand, it does not permit the modification of the interior of an elementary surface without making a corresponding change to its boundaries, which affects one or more of the neighbouring elementary surfaces. If, nevertheless, this type of modification is desired, the initial elementary surface ABCD must be divided into two by a line EF (Figure 5.61), and the patches AEFD and EBCF must be treated separately.

As a result, all the patches situated in the row bounded by the curves of the H family must be divided in two.

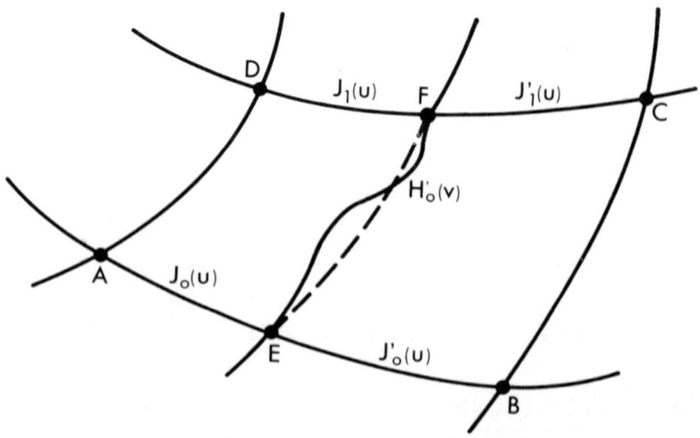

Fig. 5.61

Boundary curves. There are no restrictions on the definition of the curves H and J provided that they are expressed as a function of a parameter varying from 0 to 1 along the segment used. The use of mathematical curves raises no great difficulty, but the form of J. Ferguson's parametric curves (see p. 113) or functions of the UNISURF type (see p. 113) makes them particularly suitable.

Further information on Coons surfaces is given in Appendix 2.

The A.E.G. method

The German company A.E.G. has studied a method which allows the quasiautomatic definition of the lines of a grid situated on an arbitrary surface.

1. The coordinates of a certain number of close-spaced points are measured; for example, for the shape of a car body in 1:1 scale, their density is about 1 cm^{-2}.
2. Starting from an arbitrary point, the nearest point is searched for. Step by step, triangles are constructed with sides of minimum length, covering the entire surface. If n is the number of points, we have, for large n, about $2n$ triangles.
3. We compute ΔS, the distance between the centres of gravity of two adjacent triangles, and $\Delta\phi$, the angle between the normals to the planes of two adjacent triangles.

4. Starting from an arbitrary point (which can be the same point as in procedure 2, we determine, step by step, two chains of points such that we have $\Delta\phi/\Delta S$ maximum for one and $\Delta\phi/\Delta S$ minimum for the other.
5. Points are chosen on these two chains, and from them we determine chains having the same properties as before.
6. These chains of points are smoothed, producing a mesh of curves situated very close to the initial surface and which are approximately its geodesics and asymptotics.
7. Using this mesh of curves, we interpolate by a method similar to that of Professor Coons.

All these calculations are performed by a computer.

Biparametric definitions

J. Ferguson's method

If an overall surface is divided into elementary surfaces by curves defined as on p. 113, the curve segments AB, DC, AD and BC are defined by their end points and their end parametric derivatives, those of the first pair being taken with respect to u, and those of the second pair with respect to a second parameter v.

Let us call the ends of the derivative vectors E, F, G and H (Figure 5.62).

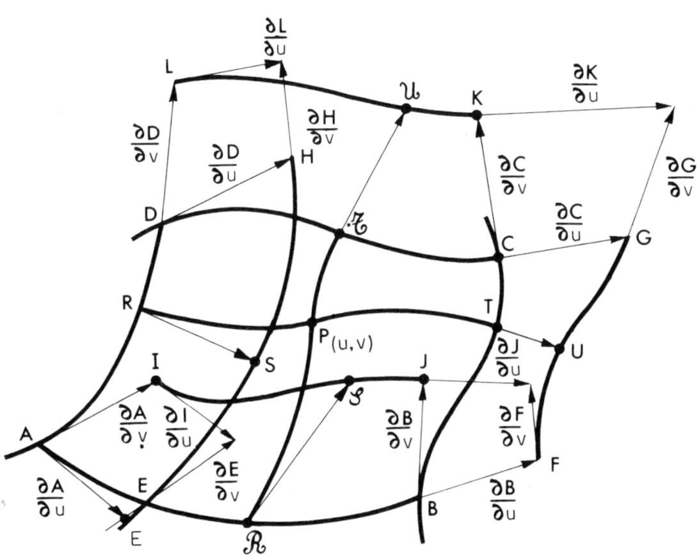

Fig. 5.62

Let us define two curves EH and FG by arbitrarily fixing four vectors:

$$\frac{\partial \mathbf{E}}{\partial v}, \frac{\partial \mathbf{F}}{\partial v}, \frac{\partial \mathbf{G}}{\partial v} \quad \text{and} \quad \frac{\partial \mathbf{H}}{\partial v}.$$

Each of the curves AD, EH, BC and FG is defined (see p. 113) by four vectors:

$$\begin{array}{cccc} \mathbf{m}_{\mathrm{AD}}, & \mathbf{n}_{\mathrm{AD}}, & \mathbf{p}_{\mathrm{AD}}, & \mathbf{q}_{\mathrm{AD}}, \\ \mathbf{m}_{\mathrm{BH}}, & \ldots, & \ldots, & \ldots, \\ \ldots, & \ldots, & \ldots, & \ldots, \\ \ldots, & \ldots, & \ldots, & \mathbf{q}_{\mathrm{FG}}. \end{array}$$

On these curves a parameter v determines points R, S, T and U such that:

$$\begin{aligned} \mathbf{R}_{(v)} &= \mathbf{m}_{\mathrm{AD}} \cdot v^3 + \mathbf{n}_{\mathrm{AD}} \cdot v^2 + \mathbf{p}_{\mathrm{AD}} \cdot v + \mathbf{q}_{\mathrm{AD}} \\ \mathbf{S}_{(v)} &= \mathbf{m}_{\mathrm{EH}} \cdot v^3 + \cdots + \cdots + \cdots \\ \mathbf{T}_{(v)} &= \cdots + \cdots + \cdots + \cdots \\ \mathbf{U}_{(v)} &= \cdots + \cdots + \cdots + \mathbf{q}_{\mathrm{FG}}. \end{aligned}$$

The generatrix $\widetilde{\mathbf{RT}}$ of the surface ABCD is defined by:

1. Its end points **R** and **T**.
2. Its parametric derivatives at these points which are respectively equal to

$$\frac{\partial \mathbf{R}}{\partial u} = \mathbf{S}_{(v)} - \mathbf{R}_{(v)} \quad \text{and} \quad \frac{\partial \mathbf{T}}{\partial u} = \mathbf{U}_{(v)} - \mathbf{T}_{(v)}.$$

Thus the surface ABCD is completely defined when we have fixed, for each of the vertices:

1. The coordinates of the vertex = **A**.
2. Three parametric derivative vectors

$$\frac{\partial \mathrm{A}}{\partial u}, \frac{\partial \mathrm{A}}{\partial v} \quad \text{and} \quad \frac{\partial \mathrm{E}}{\partial v}.$$

Conjugate generatrices. Let us call I, J, K, L the ends of the vectors

$$\frac{\partial \mathbf{A}}{\partial v}, \frac{\partial \mathbf{B}}{\partial v}, \frac{\partial \mathbf{C}}{\partial v} \quad \text{and} \quad \frac{\partial \mathbf{D}}{\partial v}.$$

Let us define two curves IJ and LK by their end points and end parametric derivatives, respectively, equal to:

$$\left(-\frac{\partial \mathbf{A}}{\partial v} + \frac{\partial \mathbf{A}}{\partial u} + \frac{\partial \mathbf{E}}{\partial v}\right), \left(-\frac{\partial \mathbf{B}}{\partial v} + \frac{\partial \mathbf{B}}{\partial u} + \frac{\partial \mathbf{F}}{\partial v}\right)$$

and

$$\left(-\frac{\partial \mathbf{D}}{\partial v}+\frac{\partial \mathbf{D}}{\partial u}+\frac{\partial \mathbf{H}}{\partial v}\right), \left(-\frac{\partial \mathbf{C}}{\partial v}+\frac{\partial \mathbf{C}}{\partial u}+\frac{\partial \mathbf{G}}{\partial v}\right).$$

On the curves AB, IJ, DC and LK a parameter u defines four points $\mathscr{R}$, $\mathscr{S}$, $\mathscr{T}$ and $\mathscr{U}$.

It can be shown* that the curves $\mathscr{R}\mathscr{T}$ defined by parametric derivatives equal to $(\mathscr{S}_{(u)}-\mathscr{R}_{(u)})$ and $(\mathscr{U}_{(u)}\mathscr{T}_{(u)})$ constitute a family conjugate to the family of curves RT: every curve of a family intersects every curve of the other family.

These two families of curves thus make up a system of biparametric coordinates on the surface ABCD. The point $P(u, v)$ at the intersection of $RT_{(u)}$ and of $\mathscr{R}\mathscr{T}_{(v)}$ has u and v for its parametric coordinates.

A relation $F(u, v) = 0$ thus represents a curve on the surface ABCD.

Continuity of curves and surfaces. For two consecutive curves to be tangential, it is necessary and sufficient that their derivative vectors at the point of contact should be collinear.

To ensure the continuity of two adjacent surfaces, Ferguson chose to use the same vectors to define the end conditions of one curve segment and the initial conditions of the next. Thus a curve $RT_{(v)}$ of one surface Σ and a curve $R'T'_{(v)}$ of a surface Σ' are continuous (Figure 5.63) since $\mathbf{R}_{(v)} = \mathbf{T}'_{(v)}$ and moreover, as

$$\frac{\partial \mathbf{R}}{\partial u} = \frac{\partial \mathbf{T}'}{\partial u} = (\mathbf{S}-\mathbf{R})$$

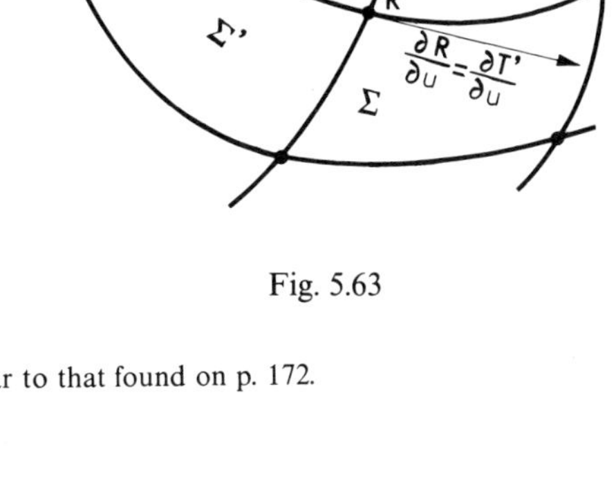

Fig. 5.63

* This proof is similar to that found on p. 172.

the curves $\widetilde{R'T'}$ and $\widetilde{RT}$ are tangential at R. The tangent planes of the two surfaces have a second common direction, since

$$\frac{\partial \mathbf{R}}{\partial v} = \frac{\partial \mathbf{T}'}{\partial v}$$

they are thus coincident, ensuring the tangency of the two surfaces at R.

This convention does not leave the designer free to fix the value of the tangent-vector magnitude at the initial point. This can be awkward if relatively contorted surfaces are to be defined, or if the adjacent patches have very different dimensions.

The UNISURF method

Principle. A surface is considered as the geometrical locus of a generatrix which undergoes simultaneously a displacement and a deformation.

The trajectories of the vertices of the characteristic polygon of the generatrix of a surface (see p. 115) are themselves canonical curves defined by their own characteristic polygons (Figure 5.64).

To a value u of the parameter there corresponds a position of each of the vertices. The characteristic polygon thus determined is that of a generatrix of the surface.

The initial generatrix G(0) is defined by a characteristic polygon $(\boldsymbol{\alpha}_1, \ldots, \boldsymbol{\alpha}_j, \ldots, \boldsymbol{\alpha}_p)$ whose vertices are $A_{0_1}, A_{0_2}, \ldots, A_{0_j}, A_{0_{j+1}}, \ldots, A_{0_p}, A_{0_{p+1}}$.

The characteristic polygons of the trajectories of the vertices are composed of the vectors $\mathbf{a}_{i_j}$ with double subscripts.

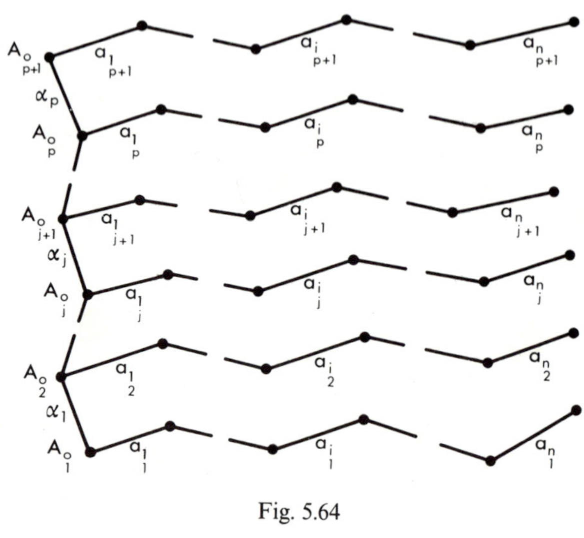

Fig. 5.64

The ensemble of the polygon $\boldsymbol{\alpha}$ and the polygons $\mathbf{a}$ constitutes the characteristic 'mesh' of the elementary surface bounded by four curves having, respectively, characteristic polygons composed of the vectors $\boldsymbol{\alpha}_i$, $\underset{1}{\mathbf{a}_i}$, $\underset{p+1}{\mathbf{a}_i}$ and the polygon having for its vertices the ends of the vectors $\underset{j}{\mathbf{a}_n}$ (Figure 5.64).

The generatrices rest on the curves defined by the polygons $(\underset{1}{\mathbf{a}_i})$ and $(\underset{p+1}{\mathbf{a}_i})$, but those which correspond to the polygons $(\underset{2}{\mathbf{a}_i}), \ldots, (\underset{p}{\mathbf{a}_i})$ do not lie on the surface itself, although they influence its definition. For this reason, the first curves are called principal directrices and the others auxiliary directrices.

All the polygons (**a**) have the same number of sides, real or zero, but it is not necessary for them to have the same number of sides as the polygon ($\boldsymbol{\alpha}$). The choice of families of functions (f) (see p. 116) depends only on the number of sides of each family of polygons.

Variable generatrix. On the principal and auxiliary directrices, a parameter u defines some points $\underset{1}{\mathrm{A}_u}, \ldots, \underset{j}{\mathrm{A}_u}, \ldots, \underset{p+1}{\mathrm{A}_u}$, which are the vertices of the characteristic polygon of the variable generatrix G_u.

These points are expressed:

$$\left.\begin{array}{l} \underset{1}{\mathbf{A}_u} = \underset{1}{\mathbf{A}_0} + \sum\limits_{1}^{n} \underset{1}{\mathbf{a}_i} \cdot \underset{n}{f_i(u)} \\ \cdots\cdots\cdots\cdots\cdots\cdots\cdots \\ \underset{j}{\mathbf{A}_u} = \underset{j}{\mathbf{A}_0} + \sum\limits_{1}^{n} \underset{j}{\mathbf{a}_i} \cdot \underset{n}{f_i(u)} \\ \cdots\cdots\cdots\cdots\cdots\cdots\cdots \\ \underset{p+1}{\mathbf{A}_u} = \underset{p+1}{\mathbf{A}_0} + \sum\limits_{1}^{n} \underset{p+1}{\mathbf{a}_i} \cdot \underset{n}{f_i(u)} \end{array}\right\} \qquad (5.73)$$

We notice that:

$$\underset{1}{\mathbf{A}_0} = \mathbf{a}_0$$

$$\underset{j}{\mathbf{A}_0} = \underset{1}{\mathbf{A}_0} + \sum_{1}^{j-1} \boldsymbol{\alpha}_j.$$

Therefore:

$$\underset{p+1}{\mathbf{A}_0} = \underset{1}{\mathbf{A}_0} + \sum_{1}^{p} \boldsymbol{\alpha}_j.$$

Variable point. On a generatrix thus determined, a parameter v defines a point $\mathrm{P}(u, v)$ which is expressed by:

$$\mathbf{P}(u, v) = \underset{1}{\mathbf{A}_u} + \sum_{j=1}^{p} (\underset{j+1}{\mathbf{A}_u} - \underset{j}{\mathbf{A}_u}) \cdot \underset{p}{f_j(v)}. \qquad (5.74)$$

Equations 5.73 and 5.74 may be expanded:

$$\underset{j+1}{\mathbf{A}_u} - \underset{j}{\mathbf{A}_u} = \underset{j+1}{\mathbf{A}_0} + \sum_{i=1}^{n} \underset{j+1}{\mathbf{a}_i} \cdot \underset{n}{f_i(u)}$$

$$- \left[\underset{j}{\mathbf{A}_0} + \sum_{i=1}^{n} \underset{j}{\mathbf{a}_i} \cdot \underset{n}{f_i(u)} \right]$$

$$= \underbrace{(\underset{j+1}{\mathbf{A}_0} - \underset{j}{\mathbf{A}_0})} + \sum_{i=1}^{n} (\underset{j+1}{\mathbf{a}_i} - \underset{j}{\mathbf{a}_i}) \cdot \underset{n}{f_i(u)}$$

$$= \boldsymbol{\alpha}_j + \sum_{i=1}^{n} (\underset{j+1}{\mathbf{a}_i} - \underset{j}{\mathbf{a}_i}) \cdot \underset{n}{f_i(u)}$$

and

$$\mathbf{P}(u, v) = \mathbf{a}_0 + \sum_{i=1}^{n} \underset{1}{\mathbf{a}_i} \cdot \underset{n}{f_i(u)} + \sum_{j=1}^{p} \boldsymbol{\alpha}_j \cdot \underset{p}{f_j(v)}$$

$$+ \sum_{i=1}^{n} \sum_{j=1}^{p} (\underset{j+1}{\mathbf{a}_i} - \underset{j}{\mathbf{a}_i}) \cdot \underset{n}{f_i(u)} \cdot \underset{p}{f_j(v)}. \tag{5.75}$$

Conjugate generatrices. Given the mesh of Figure 5.64, there exists a second mesh having the same vertices (Figure 5.65), but composed of:

The polygon of the vectors $\underset{1}{\mathbf{a}_i}$.

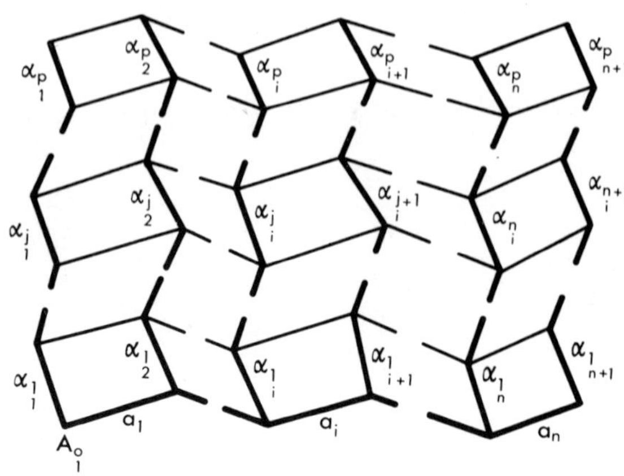

Fig. 5.65

The polygon of the vectors $\boldsymbol{\alpha}_j$.

The polygons of the vectors joining the ends of the vectors **a** with the same upper subscript.

The polygon ($\mathbf{a}$) needs only a simple subscript and the polygons ($\boldsymbol{\alpha}$) require a double subscript. On the elementary surface thus defined, the polygons composed of the segments $(\boldsymbol{\alpha}_{\substack{j\\1}})$ and $(\boldsymbol{\alpha}_{\substack{j\\n+1}})$ define the principal directrices, and the polygons $(\boldsymbol{\alpha}_{\substack{j\\2}}), \ldots, (\boldsymbol{\alpha}_{\substack{j\\n}})$ define the auxiliary directrices. As for the principal generatrices, they have, for characteristic polygons, the polygons $(\mathbf{a}_{\substack{i\\1}})$ and $(\mathbf{a}_{\substack{i\\p+1}})$ of Figure 5.64.

A parameter v defines a variable generatrix $\Gamma_{(v)}$.

On $\Gamma_{(v)}$, a point Π is defined by a parameter u. Its equation is:

$$\mathbf{\Pi}(v, u) = \mathbf{a}_0 + \sum_{i=1}^{n} \mathbf{a}_i \,.\, f_{\substack{i\\n}}(u) + \sum_{j=1}^{p} \boldsymbol{\alpha}_{\substack{j\\1}} \,.\, f_{\substack{j\\p}}(v)$$

$$+ \sum_{i=1}^{n} \sum_{j=1}^{p} (\boldsymbol{\alpha}_{\substack{j\\i+1}} - \boldsymbol{\alpha}_{\substack{j\\i}}) \,.\, f_{\substack{i\\n}}(u) \,.\, f_{\substack{j\\p}}(v). \tag{5.76}$$

If we compare this expression with equation 5.75, and if we notice (Figure 5.66) that:

$$\mathbf{a}_{\substack{i\\j+1}} - \mathbf{a}_{\substack{i\\j}} = \boldsymbol{\alpha}_{\substack{j\\i+1}} - \boldsymbol{\alpha}_{\substack{j\\i}}$$

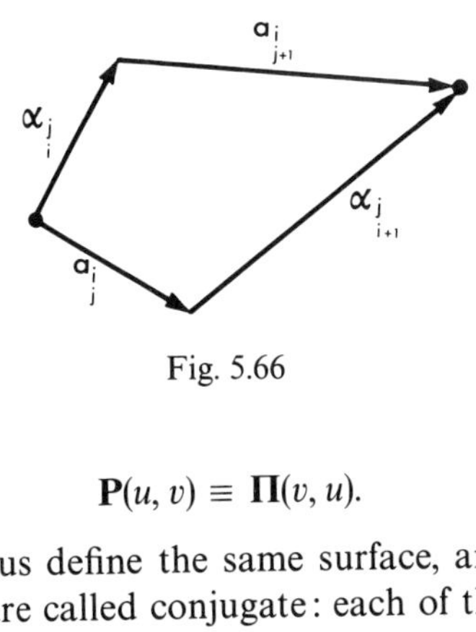

Fig. 5.66

we see that

$$\mathbf{P}(u, v) \equiv \mathbf{\Pi}(v, u).$$

The two meshes thus define the same surface, and the two families of generatrices G and Γ are called conjugate: each of them meets all those of the other family.

These two families of generatrices make up a system of biparametric coordinates on the surface thus defined.

Transposed curves. Each relationship $F(u, v) = 0$ defines, on the biparametric surface a curve called the transposed curve of the Cartesian curve,

$$F(x, y) = 0$$

provided that u and v lie within the interval (0, 1) (Figure 5.67).

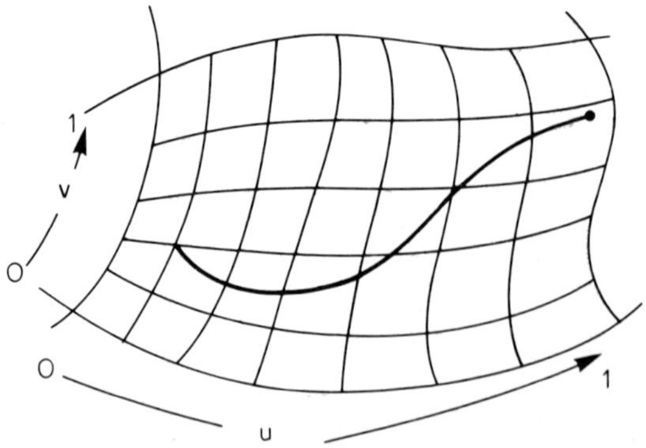

Fig. 5.67

In particular, the Cartesian curve may be defined, inside a unit patch, by a characteristic polygon and the corresponding functions.

In this case, the transposed curve is itself a canonical curve.

In effect, we have (Figure 5.68)

$$u = \sum_{0}^{n} x_k s^k$$

and

$$v = \sum_{0}^{n} y_k s^k$$

s being an auxiliary parameter expressing u and v.

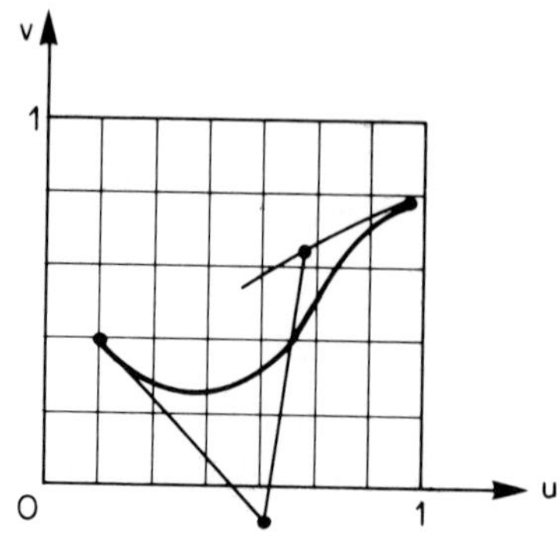

Fig. 5.68

A point $\mathbf{P}(u, v)$ is defined by

$$\mathbf{P}(u, v) = \mathbf{a}_0 + \sum_{0}^{n} \mathbf{a}_{i_{1}} f_i(u) + \sum_{1}^{p} \boldsymbol{\alpha}_i f_{i_{p}}(v) + \sum_{1,1}^{n,p} (\mathbf{a}_{i_{j+1}} - \mathbf{a}_{i_{j}}) f_{i_{n}}(u) f_{j_{p}}(v)$$

which is equivalent to:

$$\mathbf{P}(u, v) = \mathbf{a}_0 + \sum_{0}^{n} \mathbf{b}_i u^i + \sum_{1}^{b} \boldsymbol{\beta}_i v^i + \sum_{1,1}^{n,p} \gamma_{ij} u^i v^j .$$

Expressing u and v as functions of s, we find

$$\mathbf{P}(u, v) = \mathbf{P}(s) = \mathbf{a}_0 + \sum_{0}^{n} \mathbf{b}_i \left(\sum_{0}^{q} x_k s^k \right)^i + \sum_{0}^{p} \boldsymbol{\beta}_i \left(\sum_{0}^{q} y_k s^k \right)^i$$

$$+ \sum_{1,1}^{n,p} \gamma_{ij} \left(\sum_{0}^{q} x_k s^k \right)^i \left(\sum_{0}^{q} y_k s^k \right)^j .$$

The degrees, with respect to s, of the terms of this expression, are, respectively, $0, nq, pq$ and $(n+p)q$.

It is thus equivalent to

$$\mathbf{P}(u, v) = \mathbf{P}(s) = \sum_{0}^{(n+p)q} \boldsymbol{\delta}_i s^i$$

which is a canonical curve.

Continuity

Continuity of two curves. Let us consider two curves C and C′ whose initial point is $\mathbf{P}(0) \equiv \mathbf{P}'(0)$, and whose characteristic polygons are (Figure 5.69):

$$(\mathbf{a}_1, \mathbf{a}_2, \ldots \text{etc.}) \quad \text{and} \quad (\mathbf{a}'_1, \mathbf{a}'_2, \ldots \text{etc.}).$$

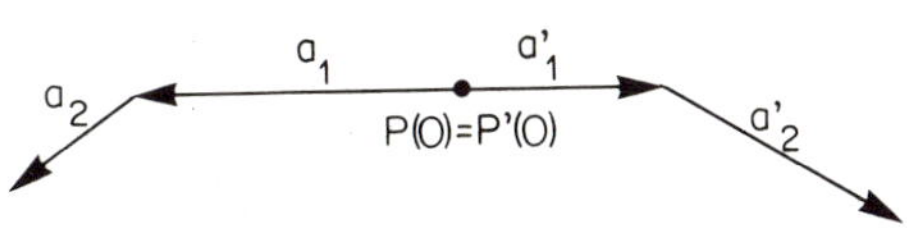

Fig. 5.69

Tangency. For C and C′ to be tangents at $\mathbf{P}(0)$, it is sufficient that the vectors $\mathbf{a}_1$ and $\mathbf{a}'_1$ should be collinear. In practice, it is also necessary for them to be in opposite directions.

The condition of tangency is thus:

$$\mathbf{a}'_1 = g \,.\, \mathbf{a}_1 \qquad (5.77)$$

$$g < 0.$$

Osculation. For the two curves C and C′ to be osculatory at $\mathbf{P}(0)$, two additional conditions must be satisfied:

(a) coincidence of the osculating planes or collinearity of the binormals
(b) identity of the radii of curvature.

As the binormals $\mathbf{b}$ and $\mathbf{b}'$ are, respectively, parallel to

$$\mathbf{a}_1 \wedge \mathbf{a}_2 \quad \text{and} \quad \mathbf{a}'_1 \wedge \mathbf{a}'_2$$

the vectors $\mathbf{a}_1, \mathbf{a}_2, \mathbf{a}'_1$ and $\mathbf{a}'_2$ must be coplanar, i.e.

$$\mathbf{a}'_2 = h\,.\,\mathbf{a}_1 + k\,.\,\mathbf{a}_2. \tag{5.78}$$

Moreover, for the principal normals $\mathbf{N}$ and $\mathbf{N}'$ to be oriented in the same direction, and not in opposite directions, we must have:

$$k > 0.$$

The curves C and C′ being defined, respectively, by polygons with n and n' sides, we have:

$$\frac{d\mathbf{P}}{du} = n\,.\,\mathbf{a}_1 \qquad \frac{d\mathbf{P}'}{du} = n'\,.\,\mathbf{a}'_1 = n'\,.\,g\,.\,\mathbf{a}_1$$

$$\frac{d\mathbf{P}}{du} \wedge \frac{d^2\mathbf{P}}{du^2} = n\,.\,\mathbf{a}_1 \wedge n(n-1)\mathbf{a}_2 = n^2(n-1)(\mathbf{a}_1 \wedge \mathbf{a}_2)$$

$$\frac{d\mathbf{P}'}{du} \wedge \frac{d^2\mathbf{P}'}{du^2} = n'^2(n'-1)\,.\,gk(\mathbf{a}_1 \wedge \mathbf{a}_2)$$

$$R = \frac{n^3|\mathbf{a}_1|^3}{n^2(n-1)|\mathbf{a}_1 \wedge \mathbf{a}_2|} = \frac{n}{n-1} \times \frac{|\mathbf{a}_1|^3}{|\boldsymbol{\alpha}_1 \wedge \mathbf{a}_2|}$$

and

$$R' = \frac{n'^3|g|^3|\mathbf{a}_1|^3}{n'^2(n'-1)|gk||\mathbf{a}_1 \wedge \mathbf{a}_2|} = \frac{n'g^2}{(n'-1)k} \times \frac{|\mathbf{a}_1|^3}{|\mathbf{a}_1 \wedge \mathbf{a}_2|}.$$

It is essential that $R = R'$. Therefore:

$$\frac{n}{n'} \times \frac{n'-1}{n-1} \times \frac{k}{g^2} = 1$$

$$g^2 = k\,.\,\frac{n}{n'}\,.\,\frac{n'-1}{n-1}. \tag{5.79}$$

For two curves to be osculatory, it is necessary and sufficient that conditions 5.77 to 5.79 should hold.

Continuity of biparametric surfaces. Notice, firstly, that since the expression 5.75 is always differentiable with respect to u and v, then, consequently, each point $\mathbf{P}(u, v)$, the conjugate generatrices $G(u)$ and $\Gamma(v)$ each have a tangent. These tangents are distinct if

$$\frac{\partial \mathbf{P}}{\partial u} \wedge \frac{\partial \mathbf{P}}{\partial v} \neq 0$$

and in this case there is a plane which is a tangent to the surface at $P(u, v)$.

Let two biparametric surfaces S and S′ be tangential, but not necessarily osculatory, along a boundary which is a principal directrix D for both (Figure 5.70).

If the auxiliary directrices are Δ and Δ', we denote by P, Q and Q′ three points defined on these curves by the same value of the parameter u.

We know that:

$$\frac{\partial \mathbf{P}}{\partial v} = n \, . \, \mathbf{PQ}$$

$$\frac{\partial \mathbf{P}'}{\partial v} = n \, . \, \mathbf{PQ}'$$

The tangency of S and S′ at P can be expressed in the general form:

$$\left(\frac{\partial \mathbf{P}}{\partial v} \wedge \frac{\partial \mathbf{P}'}{\partial v}\right) \times \frac{\partial \mathbf{P}}{\partial u} \equiv 0$$

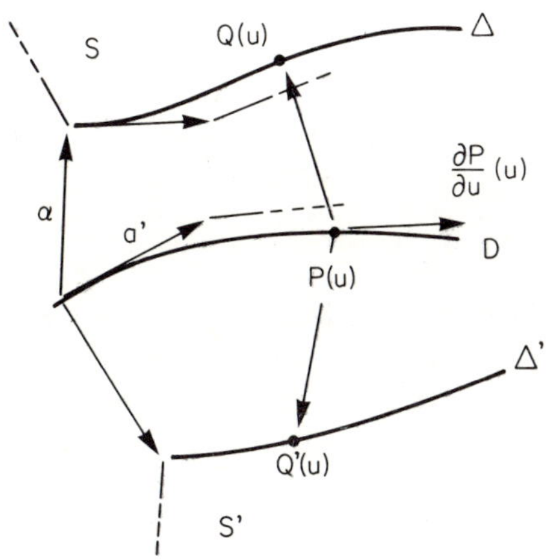

Fig. 5.70

but it is simpler to consider that it is achieved when one of these three vectors is a linear combination of the two others:

$$\mathbf{PQ}' = h(u)\mathbf{PQ}+k(u)\frac{\partial \mathbf{P}}{\partial u}(u, 0)$$

or

$$\mathbf{Q}' = \mathbf{P}+h(u)(\mathbf{Q}-\mathbf{P})+k(u)\frac{\partial \mathbf{P}}{\partial u}(u, 0) \tag{5.80}$$

$h(u)$ and $k(u)$ being scalar functions of u.

We shall express **P** and **Q** in the form:

$$\mathbf{P}(u) = \mathbf{a}_0+\sum_1^n \mathbf{b}_i \,.\, u^i$$

$$\mathbf{Q}(u) = \mathbf{a}_0+\boldsymbol{\alpha}+\sum_1^n \mathbf{c}_i \,.\, u^i$$

$$\frac{\partial \mathbf{P}}{\partial \mathbf{u}}(u) = \sum_1^n i\mathbf{b}_i \,.\, u^{i-1}.$$

Equation 5.80 takes the form

$$\mathbf{Q}' = \mathbf{a}_0+\sum_1^n \mathbf{b}_i u^i+h(u)\left[\boldsymbol{\alpha}+\sum_1^n (\mathbf{c}_i-\mathbf{b}_i) \,.\, u^i\right]+k(u) \,.\, \sum_1^n i\mathbf{b}_i u^{i-1} \tag{5.81}$$

From an examination of this relationship, we draw the following conclusions:

1. If $h(u)$ and $k(u)$ are polynomial functions, Δ' is a canonical curve.
2. If q and s are, respectively, the orders of the functions $h(u)$ and $k(u)$, the minimum number of vectors of the characteristic polygon of Δ' is equal to the greater of the two expressions

$$p+q \quad \text{or} \quad n+s-1.$$

   In particular, if $h(u)$ is a constant and $k(u)$ is a linear function, the characteristic polygon of Δ' has the same number of vectors as that of D and Δ.
3. If $h(u) =$ constant and $k(u) \equiv 0$, the vectors $\boldsymbol{\alpha}', \boldsymbol{\mu}', \mathbf{v}', \ldots, \boldsymbol{\beta}'$ are proportional to the vectors $\boldsymbol{\alpha}, \boldsymbol{\mu}, \mathbf{v}, \ldots, \boldsymbol{\beta}$, respectively, which implies that corresponding pairs are collinear.
4. If $k(u) \equiv 0$, and if the vectors $\boldsymbol{\alpha}, \boldsymbol{\mu}, \mathbf{v}, \ldots, \boldsymbol{\beta}$ are parallel, it is sufficient that $\boldsymbol{\alpha}', \boldsymbol{\mu}', \mathbf{v}', \ldots, \boldsymbol{\beta}'$ should be parallel to them, without any proportionality being imposed. In this case, a cylinder having D as a directrix and $\boldsymbol{\alpha}$ as

the direction of the generatrices is tangent to S and S′ along D; this directrix is a visible contour line of the surfaces S and S′ for a projection along the direction $\boldsymbol{\alpha}$.

Examples of the determination of the characteristic polygon of Δ'

First case: $h(u)$ = constant $= h$, $k(u) = k_0+(k_1-k_0)u$

$$\mathbf{Q}'(u) = \mathbf{a}_0+\sum_1^n \mathbf{b}_i \,.\, u^i+h\left[\boldsymbol{\alpha}+\sum_1^n (\mathbf{c}_i-\mathbf{b}_i)u^i\right]+[k_0+(k_1-k_0)u]\sum_1^n i\mathbf{b}_i u^{i-1}$$

$$= \mathbf{a}_0+h\boldsymbol{\alpha}+(1-h)\sum_1^n \mathbf{b}_i u^i+(k_1-k_0)\sum_1^n i\mathbf{b}_i u^i$$

$$+k_0\sum_1^n i\mathbf{b}_i u^{i-1}+h\sum_1^n \mathbf{c}_i u^i. \tag{5.82}$$

If we arrange that $\mathbf{b}_{n+1} = 0$, we can write

$$\sum_1^n i\mathbf{b}_i u^{i-1} = \mathbf{b}_1+\sum_1^n (i+1)\mathbf{b}_{i+1}u^i.$$

Then

$$\mathbf{Q}'(u) = \mathbf{a}_0+h\boldsymbol{\alpha}+(1-h)\sum_1^n \mathbf{b}_i \,.\, u^i+(k_1-k_0)\sum_1^n i\mathbf{b}_i u^i+k_0\left[\mathbf{b}_1+\sum_1^n (i+1)\mathbf{b}_{i+1}u^i\right]$$

$$+h\sum_1^n \mathbf{c}_i u^i.$$

Arranging this expression with respect to the powers of u, we find:

$$\mathbf{Q}'(u) = (\mathbf{a}_0+h\boldsymbol{\alpha}+k_0\mathbf{b}_1)+\sum_1^n \{[1-h+(k_1-k_0)i]\mathbf{b}_i$$

$$+k_0(i+1)\mathbf{b}_{i+1}+h\mathbf{c}_i\}u^i. \tag{5.83}$$

Equating this expression to

$$\mathbf{Q}'(u) = \mathbf{c}'_0+\sum_1^n \mathbf{c}'_i \,.\, u^i.$$

we obtain the vectors $\mathbf{c}'_i$; we find the sides of the characteristic polygon of Δ' by the operation:

$$||M_{a,b,n}||^{-1}\times||\mathbf{c}'_i||.$$

Second case: $h(u) = h_0+(h_1-h_0)u$ and $k(u) = k_0+(k_1-k_0)u$

We have:

$$\mathbf{Q}'(u) = \mathbf{P}(u)+[h_0+(h_1-h_0)u][\mathbf{Q}(u)-\mathbf{P}(u)]+[k_0+(k_1-k_0)u]\,.\,\frac{\partial \mathbf{P}}{\partial u}(u)$$

$$= \mathbf{a}_0+\sum_1^n \mathbf{b}_i u^i+[h_0+(h_1-h_0)u]\left[\boldsymbol{\alpha}+\sum_1^n(\mathbf{c}_i-\mathbf{b}_i)u^i\right]$$

$$+[k_0+(k_1-k_0)u]\sum_1^n i\mathbf{b}_i\,.\,u^{i-1}$$

$$= \mathbf{a}_0+h_0\boldsymbol{\alpha}+k_0\mathbf{b}_1+\sum_1^n \mathbf{b}_i u^i+h_0\sum_1^n(\mathbf{c}_i-\mathbf{b}_i)u^i+(h_1-h_0)\boldsymbol{\alpha}\,.\,u$$

$$+(h_1-h_0)\sum_1^n(\mathbf{c}_i-\mathbf{b}_i)u^{i+1}+(k_1-k_0)\sum_1^n i\mathbf{b}_i u^i+k_0\sum_1^n(i+1)\mathbf{b}_{i+1}u^i$$

$$= \mathbf{a}_0+h_0\boldsymbol{\alpha}+k_0\mathbf{b}_1+(h_1-h_0)\boldsymbol{\alpha}u+\sum_1^n \mathbf{b}_i u^i+h_0\sum_1^n(\mathbf{c}_i-\mathbf{b}_i)u^i$$

$$+(k_1-k_0)\sum_1^n i\mathbf{b}_i u^i+k_0\sum_1^n(i+1)\mathbf{b}_{i+1}u^i+(h_1-h_0)\sum_1^n(\mathbf{c}_i-\mathbf{b}_i)u^{i+1}$$

$$= \mathbf{a}_0+h_0\boldsymbol{\alpha}+k_0\mathbf{b}_1+(h_1-h_0)\boldsymbol{\alpha}u$$

$$+\sum_1^n[\mathbf{b}_i+h_0(\mathbf{c}_i-\mathbf{b}_i)+(k_1-k_0)i\mathbf{b}_i+k_0(i+1)\mathbf{b}_{i+1}]u^i$$

$$+(h_1-h_0)\sum_1^n(\mathbf{c}_i-\mathbf{b}_i)u^{i+1}.$$

For this expression to be of order n, as with those representing D and Δ, the coefficient of the term of degree $n+1$ must be zero, that is to say that

$$(\mathbf{c}_n-\mathbf{b}_n) = 0. \tag{5.84}$$

In the particular case where $n = 3$, if we represent the characteristic polygons of D and Δ, respectively, by the vectors $\underset{1}{\mathbf{a}_i}$ and $\underset{2}{\mathbf{a}_i}$ (Figure 5.71), equation 5.84 becomes:

$$(\underset{2}{\mathbf{a}_1}-\underset{1}{\mathbf{a}_1})-2(\underset{2}{\mathbf{a}_2}-\underset{1}{\mathbf{a}_2})+(\underset{2}{\mathbf{a}_3}-\underset{1}{\mathbf{a}_3}) = 0$$

or

$$(\boldsymbol{\mu}-\boldsymbol{\alpha})+(\boldsymbol{\beta}-\boldsymbol{\nu}) = 2(\boldsymbol{\nu}-\boldsymbol{\mu}). \tag{5.85}$$

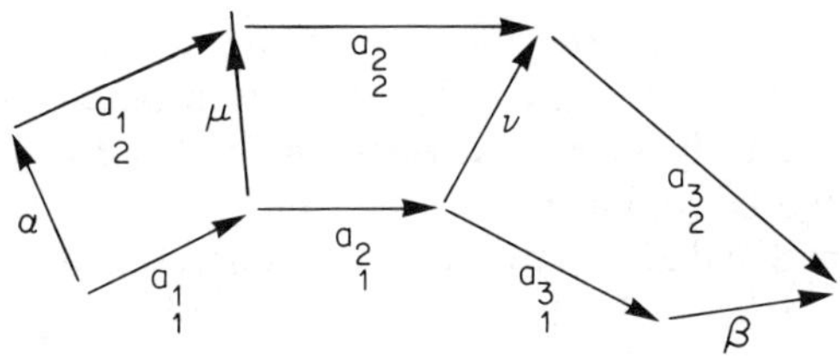

Fig. 5.71

If we draw some vectors equal to $\boldsymbol{\alpha}$, $\boldsymbol{\mu}$, $\mathbf{v}$ and $\boldsymbol{\beta}$ and having a common origin Ω, we see (Figure 5.72) that

$$\mathbf{AM} = \boldsymbol{\mu} - \boldsymbol{\alpha}$$

$$\mathbf{MN} = \mathbf{v} - \boldsymbol{\mu}$$

$$\mathbf{NB} = \boldsymbol{\beta} - \mathbf{v}.$$

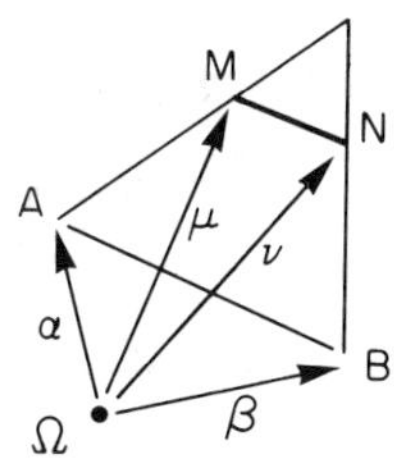

Fig. 5.72

From equation 5.85, we draw the following conclusions:

1. A, M, N and B are coplanar.
2. $\mathbf{MN} = \frac{1}{3}\mathbf{AB}$.
3. The ruled surface, whose generatrices are the straight lines joining the points having the same parameter on D and Δ, has, for its surface directrix, a cone with Ω for vertex, and the directrix defined by the polygon AMNB. This curve is a parabola, since it is represented by

$$\mathbf{P}(u) = \sum_{0}^{2} \mathbf{b}_i u^i, \qquad \mathbf{b}_3 \text{ being zero.}$$

Second-order continuity. By analogy with the results on p. 175, it would be possible to find the conditions ensuring second-order continuity between adjacent surfaces. The resulting calculations are extremely laborious, and it is simpler, in this case, to use the fifth-degree interpolation functions (see p. 165) proposed by Professor Coons.

Continuity of transposed curves. Let S and S′ be two surfaces tangent along a principal directrix D (Figure 5.73). Two transposed curves C and C′ on S and S′, respectively, meet D at the point P(u); we wish the tangents to C and C′ at P(u) to be collinear.

We transform S and S′ into two adjacent squares; C and C′ are the transposes of Γ and Γ′ (Figure 5.74).

The tangents at Π to Γ and Γ′ make the angles ψ and ψ', respectively, with the v axis*.

On the surfaces S and S′, we have (Figure 5.75):

$$\mathbf{T} \parallel \frac{\partial \mathbf{P}}{\partial u}(u, 0) \sin \psi + \frac{\partial \mathbf{P}}{\partial v}(u, 0) \cos \psi$$

or

$$\left.\begin{aligned} &\mathbf{T} \parallel \frac{\partial \mathbf{P}}{\partial u}(u, 0) \tan \psi + \frac{\partial \mathbf{P}}{\partial v}(u, 0) \\ &\text{similarly} \\ &\mathbf{T}' \parallel \frac{\partial \mathbf{P}}{\partial u}(u, 0) \tan \psi' + \frac{\partial \mathbf{P}'}{\partial v'}(u, 0). \end{aligned}\right\} \tag{5.86}$$

The continuity of T and T′ is expressed by

$$\frac{\partial \mathbf{P}}{\partial u}(u, 0) \,.\, \tan \psi + \frac{\partial \mathbf{P}}{\partial v}(u, 0) \parallel \frac{\partial \mathbf{P}}{\partial u}(u, 0) \,.\, \tan \psi' + \frac{\partial \mathbf{P}'}{\partial u'}(u, 0). \tag{5.87}$$

As the surfaces S and S′ are tangential, we have:

$$\frac{\partial \mathbf{P}'}{\partial v'} = h(u) \frac{\partial \mathbf{P}}{\partial v}(u, 0) + k(u) \frac{\partial \mathbf{P}}{\partial u}(u, 0).$$

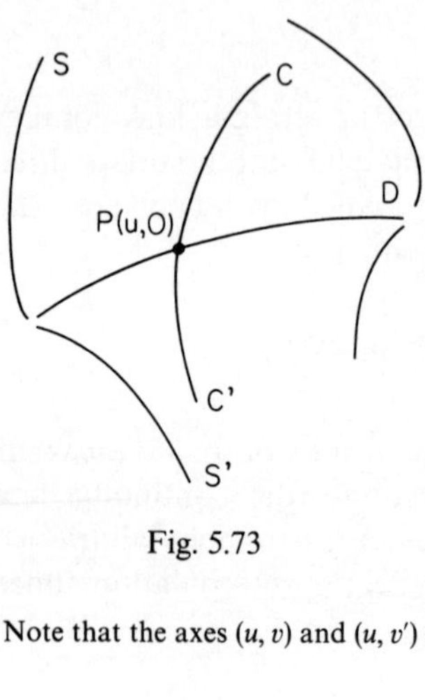

Fig. 5.73

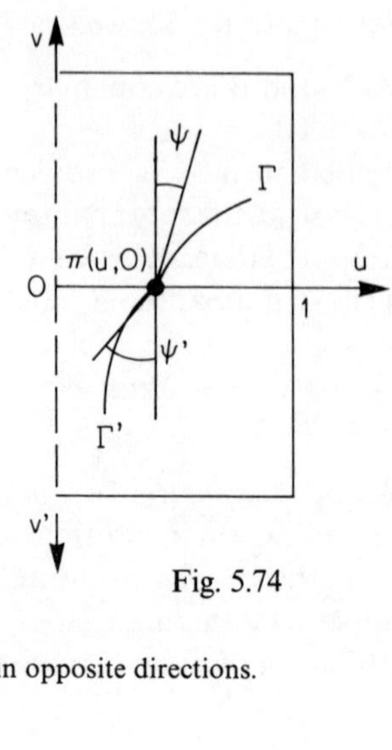

Fig. 5.74

* Note that the axes (u, v) and (u, v') are oriented in opposite directions.

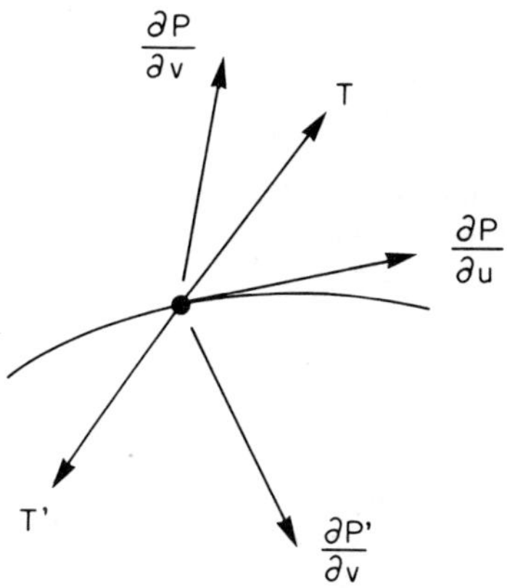

Fig. 5.75

Equation 5.87 is thus equivalent to:

$$\frac{\partial \mathbf{P}}{\partial u}(u,0)\,.\tan\psi+\frac{\partial \mathbf{P}}{\partial v}(u,0)\Big\|\frac{\partial \mathbf{P}}{\partial u}(u,0)\tan\psi'+h(u)\frac{\partial \mathbf{P}}{\partial v}(u,0)+k(u)\frac{\partial \mathbf{P}}{\partial u}(u,0)$$

or

$$\frac{\partial \mathbf{P}}{\partial u}(u,0)\,.\tan\psi+\frac{\partial \mathbf{P}}{\partial v}(u,0)\Big\|\frac{\partial \mathbf{P}}{\partial u}[k(u)+\tan\psi']+\frac{\partial \mathbf{P}}{\partial v}(u,0)\,.\,h(u)$$

whence

$$\frac{\tan\psi}{k(u)+\tan\psi'}=\frac{1}{h(u)}$$

and

$$\tan\psi' = -k(u)+h(u)\tan\psi. \tag{5.88}$$

Special cases. In an assembly of elementary surfaces, it is possible for one of them to have one or more zero sides (Figure 5.76).

The characteristic polygon corresponding to a zero side is reduced to a point coincident with the vertex. If we wish this point not to be singular, i.e. for there to be only one tangent plane to the surface, it is necessary and sufficient for the characteristic polygon of the neighbouring auxiliary directrix to be in a plane which contains the vertex coincident with the zero side.

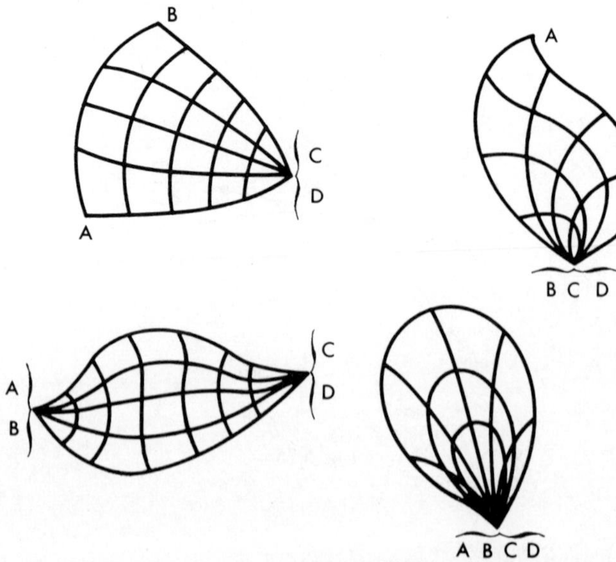

Fig. 5.76

The role of the vertices of a mesh. The end vertices of a characteristic polygon are the only points common to it and the curve it defines.

If we consider (Figure 5.77) a characteristic mesh, we see that:

1. The four corners of the mesh belong to the surface.
2. The lines defined by the polygons $(\mathbf{a}_1)$, $(\mathbf{a}_{p+1})$, $(\boldsymbol{\alpha}_1)$ and $(\boldsymbol{\alpha}_{n+1})$ lie on the surface and constitute its boundaries.
3. The envelope of the tangents to the curves of the variable generatrices along the principal directrices depends on the curves generated by the polygons $(\mathbf{a}_2)$, $(\mathbf{a}_p)$, $(\boldsymbol{\alpha}_2)$ and $(\boldsymbol{\alpha}_n)$.
4. The locus of the centres of curvature of the variable generatrices along the principal directrices depends on the curves generated by the polygons $(\mathbf{a}_3)$, $(\mathbf{a}_{p-1})$, $(\boldsymbol{\alpha}_3)$ and $(\boldsymbol{\alpha}_{n-1})$.

As a result, with a mesh having $(n+1)\times(p+1)$ vertices, we can move the $(n-3)\times(p-3)$ vertices which do not belong to the polygons $(\mathbf{a}_1)$, $(\mathbf{a}_2)$, $(\mathbf{a}_{p+1})$, $(\mathbf{a}_p)$, $(\boldsymbol{\alpha}_1)$, $(\boldsymbol{\alpha}_2)$, $(\boldsymbol{\alpha}_{n+1})$ and $(\boldsymbol{\alpha}_n)$ without altering the conditions of first-order continuity with neighbouring surfaces, and we can move the $(n-5)\times(p-5)$ points which do not belong to the polygons $(\mathbf{a}_3)$, $(\mathbf{a}_{p-1})$, $(\boldsymbol{\alpha}_3)$ and $(\boldsymbol{\alpha}_{n-1})$ without altering the conditions of second-order continuity with adjacent surfaces.

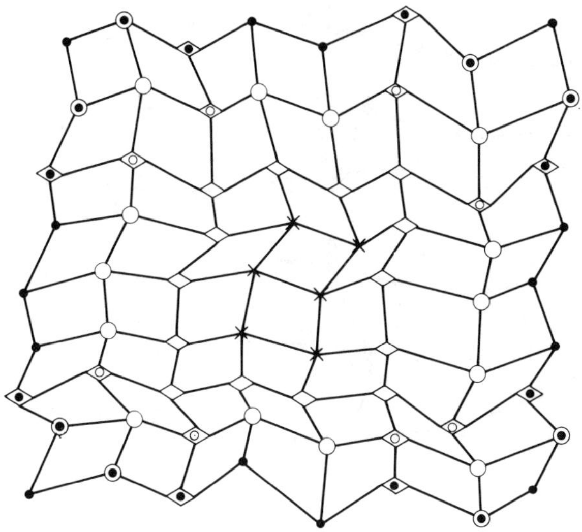

Fig. 5.77. ● = points defining the boundary curves; ○ = points defining the slope of the generatrices along the boundary curves; ◇ = points defining the curvature of the generatrices at their points situated on the boundary curves; × = points with no influence on the boundary, boundary slopes or curvatures

Generalization

Automatic definition of a surface. If a surface is defined by a sufficiently large number of points, it is possible to give an overall definition automatically:

The points are arranged on curves $C_1, C_2, \ldots, C_{n+1}$ (Figure 5.78).

Each of them, whatever the number of points lying on it ($s+1$, $t+1$, $u+1$, etc.), can be defined by a polygon with s, t, u, etc., sides.

If we consider $n+1$ points $P_{(u)}$, each of which is defined by the same parameter u on one of the curves C, we can then define the characteristic polygon of a generatrix (u) passing through all the points P(u).

Having defined m values of u between 0 and 1, we thus have m generatrices situated on the whole surface. As the characteristic polygons of the generatrices all have the same number of sides, the continuity of the surface is ensured.

In the definition of the curves C, it is possible to impose constraints represented by points through which the surface must pass, and tangents in any number whatsoever.

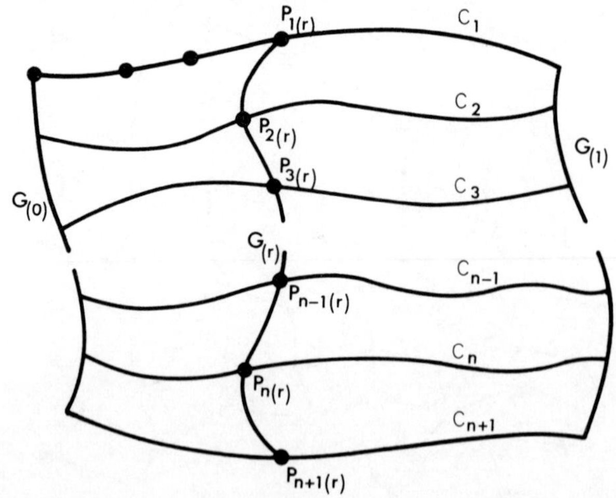

Fig. 5.78

For the characteristic polygons of the generatrices G to have the same number of sides, if the number of points is constant, the number of tangents must also be constant, or zero segments must be added. In practice, this would impose numerous computations, and it seems preferable to increase the number of points rather than to add to the constraints in the form of tangents.

Triparametric coordinates. By analogy with a characteristic mesh, it is possible to define a characteristic polyhedron made up from meshes composed of polygons with double and triple subscripts, related by a single subscripted polygon (Figure 5.79).

The points having the same coordinates (u, v) in the meshes make up a characteristic polygon defining a curve on which a point is given by a parameter w.

A vertex situated on the mesh of order k has the equation:

$$\mathbf{N}_k(u, v) = \sum_{1}^{k-1} \aleph_k + \sum_{i=1}^{i=n} \mathbf{a}_{\substack{i\\1\\k}} . f_{\substack{i\\n}}(u) + \sum_{j=1}^{j=p} \boldsymbol{\alpha}_{\substack{j\\k}} . f_{\substack{j\\p}}(v)$$

$$+ \sum_{\substack{i=1\\j=1}}^{\substack{j=p\\i=n}} (\mathbf{a}_{\substack{i\\j+1\\k}} - \mathbf{a}_{\substack{i\\j\\k}}) f_{\substack{i\\n}}(u) . f_{\substack{j\\p}}(v) \qquad (5.89)$$

A point with coordinates (u, v, w) has the equation:

$$\mathbf{P}(u, v, w) = \mathbf{N}_1(u, v) + \sum_{k=1}^{k=q} [\mathbf{N}_{k+1}(u, v) - \mathbf{N}_k(u, v)] f_{\substack{k\\q}}(w)$$

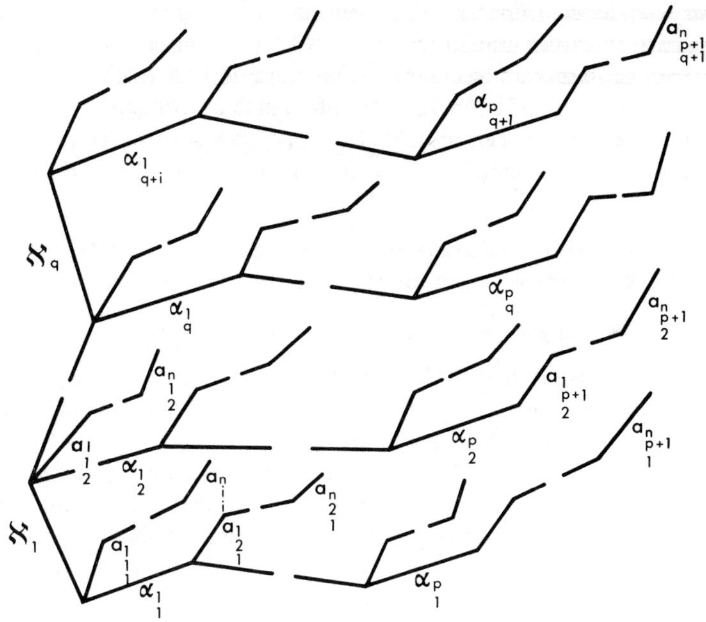

Fig. 5.79

Taking into account equation 5.89, we have:

$$\mathbf{P}(u, v, w) = \mathbf{a}_0 + \sum_{i=1}^{i=n} \mathbf{a}_{\substack{i\\1\\1}} . f_{\substack{i\\n}}(u) + \sum_{j=1}^{j=p} \boldsymbol{\alpha}_{\substack{j\\1}} . f_{\substack{j\\p}}(v) + \sum_{k=1}^{k=q} \aleph_k . f_{\substack{k\\q}}(w)$$

$$+ \sum_{\substack{i=1\\j=1}}^{\substack{j=p\\i=n}} \left(\mathbf{a}_{\substack{i\\j+1\\1}} - \mathbf{a}_{\substack{i\\j\\1}}\right) . f_{\substack{i\\n}}(u) . f_{\substack{j\\p}}(v) + \sum_{\substack{i=1\\k=1}}^{\substack{k=q\\i=n}} \left(\mathbf{a}_{\substack{i\\1\\k+1}} - \mathbf{a}_{\substack{i\\1\\k}}\right) . f_{\substack{i\\n}}(u) . f_{\substack{k\\q}}(w)$$

$$+ \sum_{\substack{j=1\\k=1}}^{\substack{k=q\\j=p}} \left(\boldsymbol{\alpha}_{\substack{j\\k+1}} - \boldsymbol{\alpha}_{\substack{j\\k}}\right) . f_{\substack{j\\p}}(u) . f_{\substack{k\\q}}(w)$$

$$+ \sum_{\substack{i=1\\j=1\\k=1}}^{\substack{k=q\\j=p\\i=n}} \left(\mathbf{a}_{\substack{i\\j+1\\k+1}} + \mathbf{a}_{\substack{i\\j\\k}} - \mathbf{a}_{\substack{i\\j+1\\k}} - \mathbf{a}_{\substack{i\\j\\k+1}}\right) . f_{\substack{i\\n}}(u) . f_{\substack{j\\p}}(v) . f_{\substack{k\\q}}(w) \qquad (5.90)$$

Each surface or curve inscribed on a unit cube and represented by a relation or group of relations between u, v and w can thus undergo a linear transformation given by equation 5.90.

Multiparametric coordinates. The concepts developed in the preceding section are applicable to abstract spaces with any number of dimensions.

It is very probably unnecessary to make use of these methods, or even of those in the preceding section, for handling current problems.

Their existence has been indicated because they seem capable of helping the representation of phenomena which are functions of more than three variables.

Descriptive operations and transposition. The methods outlined on p. 136 and p. 182 are applicable when multiparametric definitions are used.

THE TRAJECTORY OF THE POINT OF CONTACT OF A TOOL

If a surface has been defined by any one of the procedures described in the preceding sections, there remains to be determined the trajectory of the point of contact of the tool which is to cut it.

Between the surface to be cut and the envelope surface of the tool there is generally only one contact point (Figure 5.80); so that the tool cuts furrows separated by ridges which must be removed by final finishing.

The distance between the furrows is determined, taking into account the ratio of the cost of machining to the cost of finishing.

As the use of numerical control has allowed an improvement in the speed and precision of contouring machines, some users have decided to reduce the distance between finishing cuts to eliminate hand finishing more or less completely.

These remarks do not apply to cases where the contact between the surface and the tool is linear—the milling or grinding of surfaces generated by a tool with a constant profile.

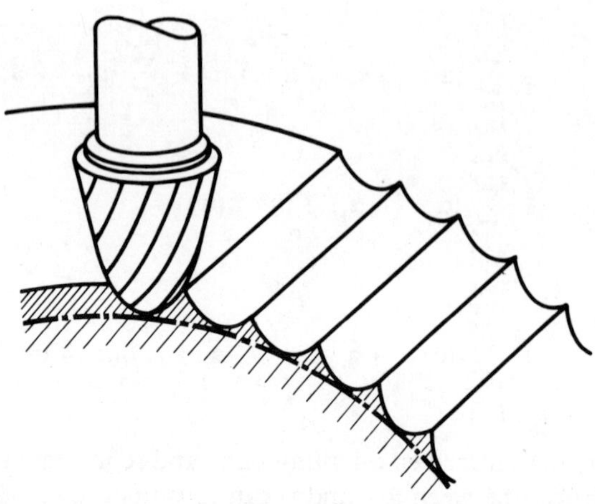

Fig. 5.80

There are several methods for determining the trajectory of the point of contact of the tool and the surface to be cut.

One of these, which is part of the A.P.T. program, may be depicted as shown in Figure 5.81.

The part surface to be cut, P, is intersected by a surface D called the drive surface.

The tool is constrained to move at a tangent to P and to D until it becomes a tangent to a third surface C, called the check surface.

This travel is defined by a trajectory KL.

Next, the preceding check surface is regarded as a drive surface D′, and a third surface C′ is used as the check surface.

When the trajectory LM has been covered, C′ becomes D″ and the check surface C″ is a surface derived from D by a simple operation such as translation.

Thus, step by step, the surface is entirely traversed.

The choice of the surfaces D, C, etc., is left to the judgement of the programmer, who takes into account the mode of definition of P. One of the simplest solutions is to define them as planes parallel to the coordinate axes.

If the surface to be cut is defined biparametrically, it is easy to make the trajectory coincide with a family of generatrices. However, this necessitates

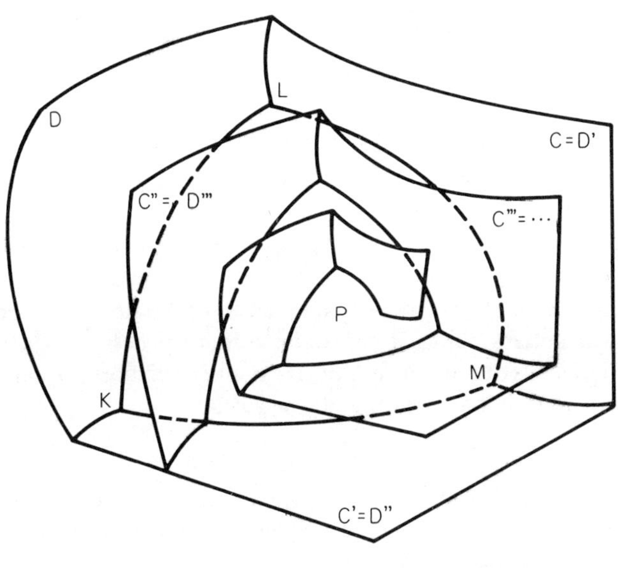

Fig. 5.81

the use of a three-dimensional control system. Moreover, the generatrices could converge and, as a result, the furrows would be very close to each other on one part of the surface, leading to a loss of time.

We see, for example, that on the same surface (Figure 5.82), the total length of the trajectories can vary by a factor of 2 to 1, according to whether path 1 or path 2 has been chosen.

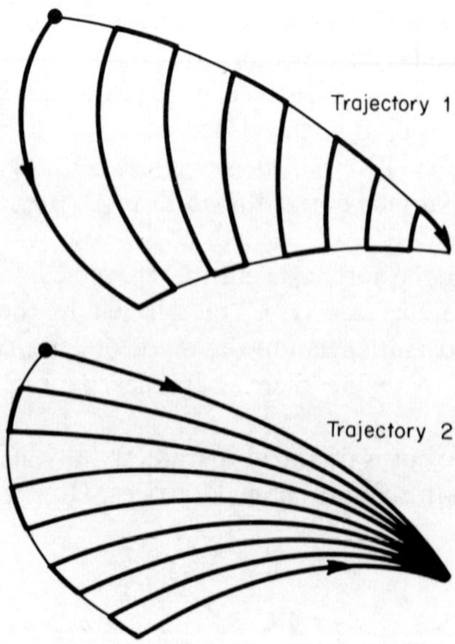

Fig. 5.82

CORRECTION FOR THE CUTTER RADIUS

If soft material (wood, plaster, plastics) is being worked, it is possible to use a point tool whose tip, lying on the axis of the spindle, describes the successive generatrices of a surface leaving furrows with triangular cross-sections.

As soon as the material to be machined offers some resistance, the use of a cylindrical, sphero-conical or toroidal cutter is necessary: the trajectory of a point on the axis of the tool spindle is no longer deduced by a translation of the point of contact of the tool with the surface.

Two-dimensional contouring

Suppose that the directrix curve of the cylindrical machining surface is situated in a plane parallel to one of the planes of the coordinate axes.

If this curve is defined by the relation:

$$F(x, y) = 0$$

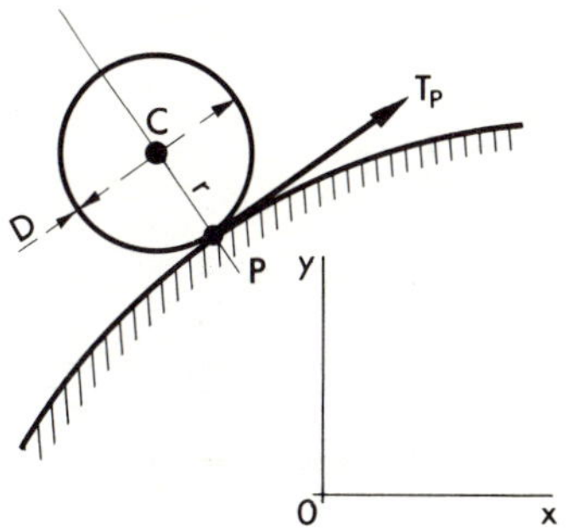

Fig. 5.83

at a point of contact P(x, y) (Figure 5.83), the centre C of the cylindrical tool of diameter D has the corresponding coordinates:

$$x_C = x_P \pm \frac{D}{2} \cdot \frac{dy}{\sqrt{dx^2+dy^2}}$$

$$y_C = y_P \pm \frac{D}{2} \cdot \frac{dx}{\sqrt{dx^2+dy^2}}.$$

The + sign or the − sign is used according to whether the tool is to the left or the right of the line to be cut.

In the Ferranti system, the diameter D is chosen by the part programmer and a tool with the prescribed diameter must be used, or a roughing cut must be performed, if necessary, with a tool of smaller diameter.

Other systems (Olivetti, for example) introduce into the block of data relative to each end of a cut the values of the direction cosines of the normal, and it is the machine's control system which makes the correction, allowing machining with any diameter of tool or the use of the same tool for roughing and finishing.

There are other control systems, in addition, which themselves calculate the direction cosines of the normals and from them deduce the correction to be applied to the coordinates, taking into account the radius of the tool.

The speeds of displacement of the centre of the tool and the point of contact along the trajectory are in the ratio (Figure 5.84)

$$\frac{V_P}{V_A} = \frac{R}{R \pm r}.$$

The + or − sign is used according to whether the surface being machined is convex or concave.

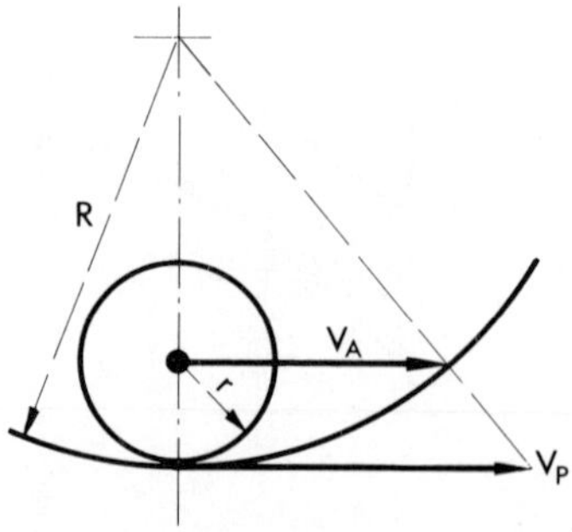

Fig. 5.84

If the control system maintains a constant value for V_A, V_P can become very large if the surface is concave and if R and r are almost identical, leading to a deterioration of the tool if the required precautions are not taken.

Three-dimensional contouring

Spherical tools

Spindle with fixed orientation

When the point of contact of a spherical cutter describes a curve C on the surface S being machined (Figure 5.85), the centre Ω of the sphere describes a trajectory D situated on a surface parallel to S separated by the distance r, the radius of the tool.

In general, the computer works out the coordinates of the points on D during the preparation phase, and we assume then that the radius of the tool is known in advance.

This hypothesis is convenient, because it is easy to grind a sphero-conical tool without modifying its geometric characteristics.

To our knowledge, there are, as yet, no three-dimensional contouring systems in which the direction cosines of the normals are written into every block of data.

The direction of the normal to a surface cannot be calculated on the spot when only one curve passing through a point is known.

If we use a biparametric definition, we calculate the two partial derivatives $\partial\mathbf{P}/\partial u$ and $\partial\mathbf{P}/\partial v$ whose vectorial product is parallel to the normal.

To obtain the points of a generatrix $\Gamma(v)$, we calculate beforehand the vectors $\underset{v}{\mathbf{a}_i}$, and the coordinates of the general point $\mathrm{P}(u, v)$ are expressed by

$$\mathbf{P}(u, v) = \mathbf{a}_0 + \sum_1^p \boldsymbol{\alpha}_i \cdot \underset{p}{f_i}(u) + \sum_1^n \underset{v}{\mathbf{a}_i} \cdot \underset{n}{f_i}(u).$$

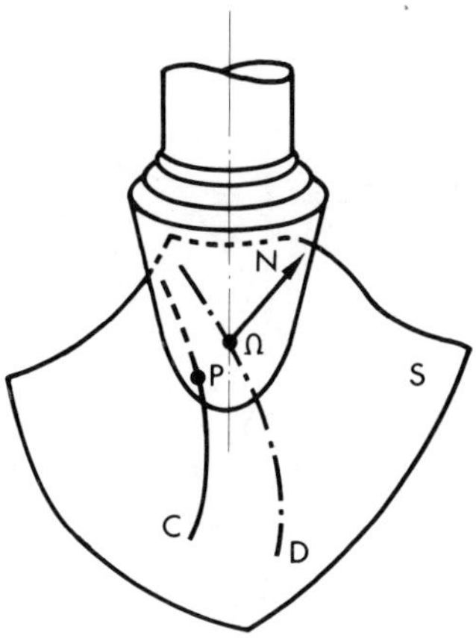

Fig. 5.85

The derivative $\partial \mathbf{P}/\partial u(u, v)$ is thus:

$$\frac{\partial \mathbf{P}}{\partial u}(u, v) = \sum_{1}^{n} \mathbf{a}_{i_v} \cdot f'_{i_n}(u)$$

To calculate $\partial \mathbf{P}/\partial v(u, v)$, we derive the expression:

$$\mathbf{P}(u, v) = \mathbf{a}_0 + \sum_{1}^{n} \mathbf{a}_{i_1} \cdot f_{i_n}(u) + \sum_{1}^{p} \boldsymbol{\alpha}_i \cdot f_{i_p}(v) + \sum_{1,1}^{n,p} (\mathbf{a}_{i_{j+1}} - \mathbf{a}_{i_j}) f_{i_n}(u) f_{j_p}(v)$$

whence

$$\frac{\partial \mathbf{P}}{\partial v}(u, v) = \sum_{1}^{p} \boldsymbol{\alpha}_i \cdot f'_{i_p}(v) + \sum_{1,1}^{n,p} (\mathbf{a}_{i_{j+1}} - \mathbf{a}_{i_j}) \cdot f'_{j_p}(v) \cdot f_{i_n}(u). \tag{5.91}$$

In the block of information necessary to compute the trajectory of a tool following the generatrix $\Gamma(v)$, we thus find, on the one hand, the values of the vectors $\mathbf{a}_i$ (or, rather, the corresponding vectors $\mathbf{b}_i$), and, on the other hand, the constants

$$\sum_{1}^{p} \boldsymbol{\alpha}_i \cdot f'_{i_p}(v) \quad \text{and} \quad (\mathbf{a}_{i_{j+1}} - \mathbf{a}_{i_j}) \cdot f'_{j_p}(v)$$

by means of which, for each point $\mathbf{P}(u, v)$, we readily compute $\partial \mathbf{P}/\partial v(u, v)$.

It is also possible to adopt an approximate solution, which consists of calculating simultaneously the coordinates of four points defined by parameters v and $v + \Delta v$, lying on two adjoining generatrices corresponding to the parameter values u and $u + \Delta u$ (Figure 5.86).

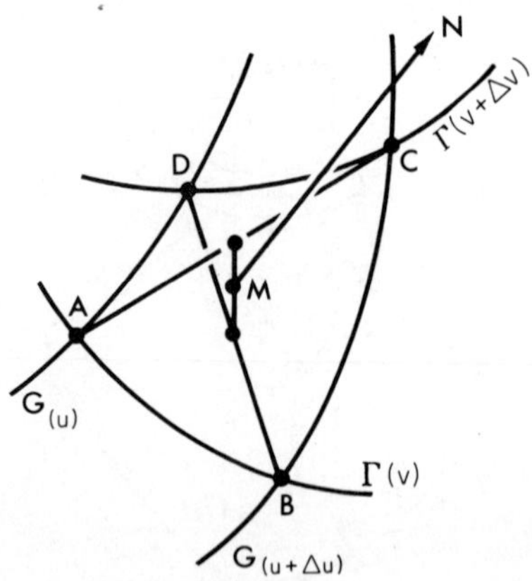

Fig. 5.86

We then assume that a point

$$\mathbf{M} = \frac{\mathbf{A}+\mathbf{B}+\mathbf{C}+\mathbf{D}}{4}$$

and the normal

$$\mathbf{N} = \pm(\mathbf{D}-\mathbf{B}) \wedge (\mathbf{C}-\mathbf{A}) \tag{5.92}$$

can be considered as an acceptable solution.

The calculation of the value of expression 5.92 is much simpler than that of expression 5.91, since no derivatives have to be computed. The sign + or − is chosen according to whether the tool, for a given direction of travel on the generatrices, is assumed to be above or below the surface.

Variable-orientation spindle

If the machine is fitted with a variable-orientation spindle, and if the angular movements of the spindle are controlled in such a fashion that its axis remains constantly normal to the surface being machined (Figure 5.87), it is sufficient, for the point P to describe the trajectory ST situated on S, that the normal should always pass through the centre of the sphere, whatever the radius of the tool.

In practice, the portions of cutting edges situated near to the axis of the spindle work in mediocre conditions, and the grinding of a 'down-to-centre' cutting tool raises some difficulties. To avoid these inconveniences, we can define a direction N which passes through the centre of the sphere and makes a constant angle with the axis B of the spindle and the axis C of its pivot (Figure 5.88).

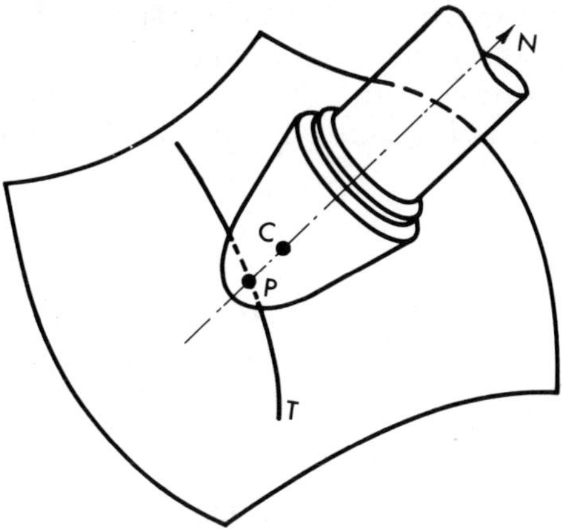

Fig. 5.87

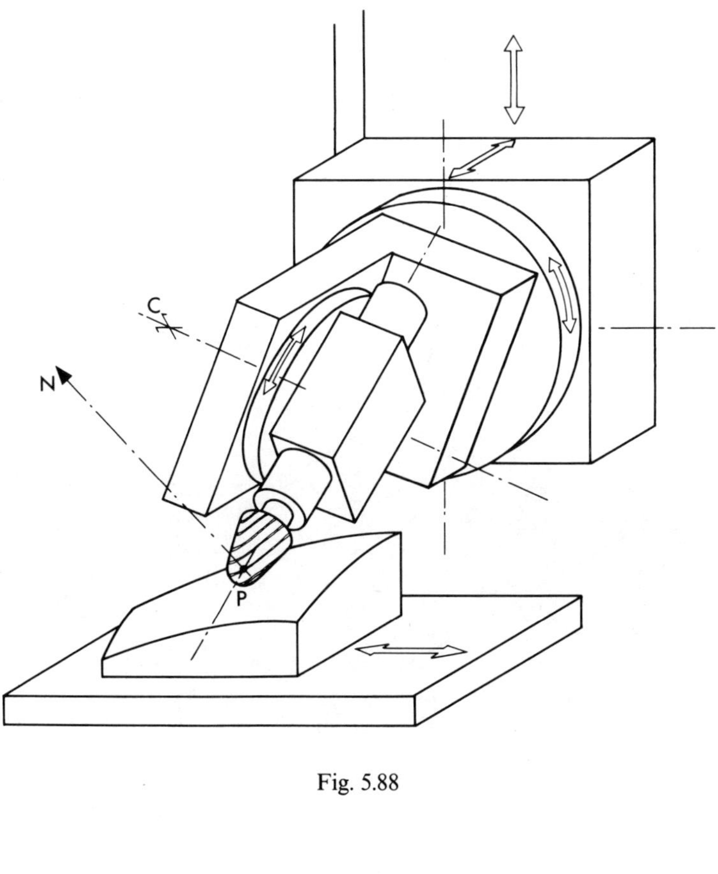

Fig. 5.88

The point of intersection (P) of N and the sphere plays the same role as the point P in Figure 5.85, and if N is constantly normal to the surface being cut, no correction need be made. The point P, not lying on the axis of the tool, has a non-zero cutting speed, allowing rapid wear at the centre to be avoided.

Toroidal tool

Cutting tools in the form of a toroid have more favourable cutting characteristics than spherical tools; therefore they are used in preference each time large quantities of material are to be removed.

Let r be the radius of the circumference which generates the torus and let R be that of the locus of its centre.

To calculate the correction, it is assumed that the axis of the torus is parallel to one of the coordinate axes, but one must be prepared to perform, if necessary, a further rotation on the data thus obtained, to place them in the coordinate system of the machine.

Let P be the point of contact between the tool and the part, C the section of the surface S in a plane perpendicular to the axis of the tool passing through P, and D the section in the plane containing P and the axis of the tool.

The direction cosines of the normal to the surface at P are represented by u, v and w.

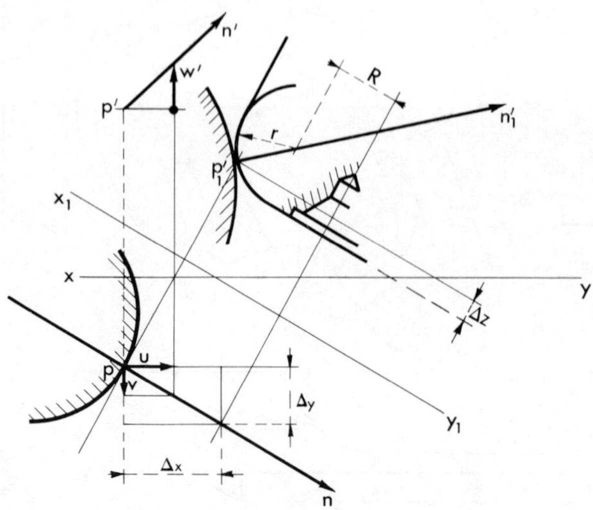

Fig. 5.89

It is seen (Figure 5.89) that the corrections to be applied to the centre point E of the tool are given by the expressions:

$$\Delta X = (R+r\sqrt{1-w^2})\frac{u}{\sqrt{u^2+v^2}}$$

$$\Delta Y = (R+r\sqrt{1-w^2})\frac{v}{\sqrt{u^2+v^2}}$$

$$\Delta Z = -r(1-w).$$

References

1. R. A. Kelley, 'APT-ADAPT', *American Machinist*, Special Report No. 554, 22 June 1964.
2. R. A. Thomas, 'The languages of tape', *American Machinist*, Special Report No. 545, 6 January 1964.
3. R. M. Sim, 'The programming of numerically controlled machines (La programmation des machines à commande numérique)', *Machinery*, 23 February 1966.
4. W. T. K. Henderson, 'The 2C,L program for numerical control', *Machinery and Production Engineering*, 18 October 1967.
5. 'EXAPT for point-to-point and simple straight-line control (EXAPT für Punkt und einfache Streckensteuerungen)', Document of the Technische Hochschule, Aachen, October 1966.
6. J. Guillemet, 'The IFAPT programming language (Le langage de programmation IFAPT)', *La Machine-Outil française*, April 1968.
7. R. M. Sim, 'Numerically-controlled machine tool programming', *Machinery and Production Engineering*, 23 February 1966.
8. J. Ponte, 'The "Profile Data" system (Le système "Profile Data")', *Mécanique et Électricité*, August 1968.
9. H. Opitz, 'Automatic programming of numerically controlled machines (Programmation automatique des machines à commande numérique)', Report of the C.I.R.P. European Conference on 24th and 25th April 1968. Published by G.A.M.I.
10. G. Sauvaire, 'The integration of technological data (Intégration des données technologiques)', *Mécanique et Électricité*, June 1968.
11. E. Merchant, 'Progress and problems (Progrès et problèmes)', *Mécanique et Électricité*, August 1968.
12. R. Weill, 'Optimization of the cycle (Optimalisation du cycle)', *G.A.M.I.*, **1966**, special issue.
13. G. Le Tac, 'The definition of curves and surfaces in the APT and ADAPT languages (Définition des courbes et surfaces en langage APT et ADAPT)', *Mécanique et Électricité*, August 1968.
14. H. Michaelis, 'Interpolation with natural polynomials (Interpolation mit natürlichen Polynomen)', *V.D.I. Zeitung*, 11 December 1962.
15. D. A. Atkins and R. A. Tapia, 'Mathematical ship lofting', Society of Naval Architects and Marine Engineers, Santa Barbara Congress, 13 October 1962.
16. J. Ferguson, 'Multivariable curve interpolation', *Journal of the Association for Computing Machinery*, April 1964.

17. P. Bézier, 'The use of numerical control in the study of car bodies and tooling (Emploi de la commande numérique pour l'étude des carrosseries et des outillages)', *Journal of the S.I.A.*, April 1968.
18. G. Fayard, 'Mechanical manufacture from mathematical equations (Les fabrications mécaniques à partir des équations mathématiques)', *La Technique Moderne*, April 1961.
19. R. Moreau, 'Mathematical machine for cutting complex shapes (Machine mathématique à usiner les formes compliquées)', *Sciences et Techniques*, May 1966.
20. S. A. Coons, 'Surfaces for computer-aided design of space forms', Technical Report MAC-TR-41, M.I.T., Cambridge, Mass., June 1967.
21. 'Advanced techniques for body engineering (GMC)', *Automotive Engineer*, March 1969.
22. 'Chevrolet is using computers, NC, EDM, abrasive machining and advanced casting techniques to reduce stamping die costs', *Machinery*, May 1969.
23. W. L. Johnson, 'Interactive computer graphics as an engineering tool', Document SAE 69 0192, Detroit congress, January 1969.
24. 'ICL off-line system for computer produced drawings', *Machinery and Production* Engineering, 4 June 1969.
25. M. Chalvet, 'A designing and sculpting machine—the Renault/CII Unisurf process (Machine à dessiner et à sculpter, procédé Unisurf Renault/CII)', *Ingénieurs et Techniciens*, June 1969.
26. 'Adaptive control for machine tools', *Machinery and Production Engineering*, 10 April 1968.
27. R. Weill, 'Future developments in numerically controlled machine tools (Développements futurs de la commande numérique des machines outils)', INSA Meeting, 1968.
28. 'Adaptive control goes into production', *Steel*, 29 April 1968.
29. 'Cincinnati Acramizer—adaptive feed and speed optimizing control', Cincinnati publication No. SP 112.
30. R. A. Mathias, 'An effective system for adaptive control of the milling process', Technical paper, ASTME, MS 68 202.
31. 'The influence of numerical control and adaptive control upon cutting tool design', *Metal cuttings*, 2nd and 3rd quarter 1968, published by the National Twist Drill and Tool Company.
32. B. Porter and R. D. M. J. Summers, 'Adaptive machine tool control, the state of the art', *Machinery and Production Engineering*, 5 February 1969.
33. R. Khol, 'Adaptive control towards the thinking machine', *Machine Design*, 1 May 1969.
34. H. Takeyama, H. Sekiguchi and K. Takada, 'One approach for optimizing control in metal cutting', *Annals of the CIRP*, **XVII**, 1969.
35. G. Spur, 'The optimizing of the machine tool's performance', *Mécanique et Électricité*, October 1969.
36. A. de Filippi and R. Ippolito, 'Adaptive control in turning', General Assembly of the C.I.R.P., September 1968.
37. H. Hollingum, 'Sensors govern progress in adaptive control', *Metalworking Production*, 24 June 1970.

Chapter 6

Exploitation of Machines

6.1 GENERALITIES

If one compares numerically controlled machine tools with conventional machines, there is sometimes a similarity between the external shapes, but one always finds a very great difference between their functional characteristics.

A single statistic suffices to demonstrate this. In a conventional workshop, cutting time rarely exceeds 15 per cent of the total working time. During the rest of the time, the operator is reading the drawings, clamping and unclamping the part, mounting and demounting the tools, executing the displacements, measuring, setting the speeds and removing swarf. In a workshop with numerically controlled machines operating under good conditions, this figure reaches 85 per cent, because everything has been organized to eliminate dead time in the machine cycles. In particular, all the preparatory jobs are carried out when the machines are performing other operations.

When the cost is seen in this perspective, the use of numerically controlled machines, which require a greater investment than conventional machines of comparable dimensions, is nevertheless more economical.

As a result, the greatest advantage cannot be extracted from the use of numerically controlled machine tools unless the effort is made to adapt the method of working, in all activities and at all levels of a company, to the requirements of numerical control.[1–7]

6.2 CONCEPTION OF THE PRODUCT

Sometimes the use of a new technique has completely altered the appearance and fundamental characteristics of an object. A typical example is supplied by numerically controlled machine tools, some of which bear no resemblance to traditional machines, or by integrally machined aircraft structural members, which have replaced members built up by riveting parts cut from sheet or from sections.[8]

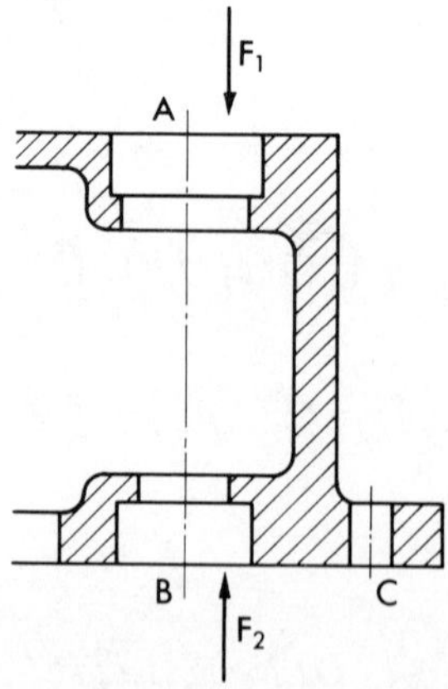

Fig. 6.1

Numerical control requires, much more frequently, a modification of the details required for manufacturing the parts which make up an assembly, or at least of those that are amenable to this technique.

As, in this case, the first consideration must be to reduce the dead time, efforts are made to design parts that are capable of being machined completely without having to be demounted and turned over on the machine table. Thus, the gearcase of Figure 6.1 must be tackled from the direction F_1 for the machining of the hole A, and from the direction F_2 for the machining of B and C.

On the other hand, the part shown in Figure 6.2 can be tackled in the direction F_1; it is sufficient that the hole B should not have a shoulder and that C should be offset from the vertical face. This implies, also, that the bearing mounted in B should be fitted with a flange or a keeper ring.

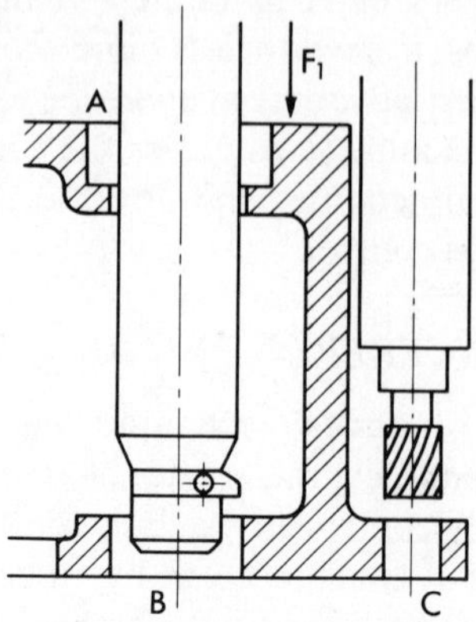

Fig. 6.2

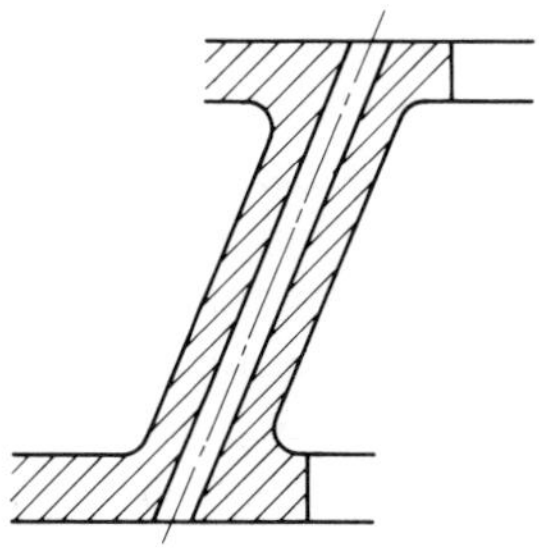

Fig. 6.3

Similarly, an oblique lubricating hole (Figure 6.3) may be replaced by three holes parallel to the coordinate axes (Figure 6.4). At the cost of a

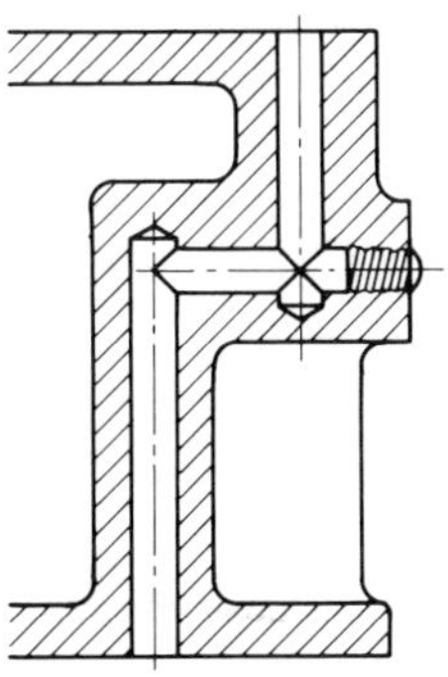

Fig. 6.4

supplementary tapping operation, this allows the use of a three-axis machine, the hourly cost of which is lower than the five-axis machine which would be necessary for drilling an oblique hole.

All these observations are valid for conventional machines, but, as numerically controlled machines have a higher hourly cost, the attention of work planning has been drawn to some points which have been overlooked too often and for too long.

6.3 DIMENSIONING OF DRAWINGS

Functional dimensioning possesses indisputable advantages, but if we wish to machine a part on a machine where the position of the moving parts is marked by a graduated scale, we will, at some time, have to transform the dimensions of the drawings into a reference system equivalent to that of the machine. Thus the values read on the machine differ from the dimensions of the drawing only by constant values representing the projections on the reference system of the machine of the vector linking it by translation to the reference system of the part dimensions.

This rule has been followed for a long time, in a very general way, for all the operations performed on jig boring machines.

For parts drilled and bored on numerically controlled machines, this description is advantageous when the program is written manually, an automatic calculator playing a more or less important role in the predetermination of the coordinates.

The use of programming languages frees the part programmer from working out the data of geometrical problems posed by the transformation of dimensions; given the facilities of the most widely used programs (see p. 99), the dimensioning of drawings can be effected almost as the designer wishes, and the transformation of the dimensions into a Cartesian frame of reference hardly poses any difficulties. Automatic programming has thus given the designer a much greater liberty in the choice of the dimensioning system, since he no longer has to make the trigonometric calculations which he was previously obliged to perform by himself. Nevertheless, the adoption of a simple and logical reference system eases the task of the programmer and reduces the risks of error.

6.4 WORK PREPARATION

Process charts

The introduction of numerical control has not brought about a fundamental change in the part played by the methods department.

It is necessary, however, to study in depth the range and nature of each operation. Indeed, if an error was made, it would hardly be noticed until work was in progress, and a modification at this time causes a serious loss of time and a disruption in the general work scheme.

Therefore the introduction of numerical control inevitably brings with it an increase in preparatory work, compensated, on the other hand, by a reduction of the demands made on the skilled and trained worker.

The program

When the sequence of operations has been established, there remain to be written the instructions which the computer requires to carry out all the logical or arithmetical operations necessary for the complete formulation of the commands to be sent to the machine.

To draw up this program, the problems which the computer can solve, and the form in which the data must be supplied to it, must be known.[9]

This knowledge is not, at present, widespread among all method planners, and it is the programmers who act as interpreters between the planners and the computer, translating the ideas of the first into a form assimilable by the second.[10–11]

The complexity of their work varies greatly, according to whether the parts are simple enough to be machined on two- or three-axis machines, or require the use of five- or six-axis machines.

For the sake of simplicity and economy, the work of the computer is divided into two phases:

1. All the calculations related to the geometry of the part and to the theoretical conditions of work are performed. The results of these operations do not depend on the type of machines to which the work is entrusted or the characteristics of their control system.
2. The results are adapted to suit to the particular machine selected for performing the work.

This is why we distinguish, during this phase of the work, between the general program and the post processor.

THE GENERAL PROGRAM

A program is composed of some information which is valid for a series of operations, other information concerning all the work to be performed, and, finally, information related to a specific operation.

The first category includes the geometric form of the tools (diameter, length, chamfer, radius, etc.) and their working conditions (speed, feed, coolant, etc.).

The second category of data defines the position of the part on the table, the space taken up by clamps, etc.

The final category contains the description of the work cycle to be accomplished, the geometrical data defining the path of the tool and, in certain cases, special working conditions (speed, precision, deceleration at the end of a cut, etc.).

Sometimes the third category is subdivided into two sections. In the first we find geometrical information by means of which the coordinates of points characterizing each of the operations are calculated. At the same time, the points or groups of points are given an alphanumeric symbol.

In the second section, the sequence of elementary operations which make up the complete program is described phase by phase.

The data related to the geometry of the part or of the tool path are expressed by combinations of letters which are abbreviations of English words or expressions. To simplify the task of the part programmers, attempts have been made to create a version derived from the appropriate language for each country. In practice, the English vocabulary is quickly learned, and the existence of equivalent versions linked to different languages can only lead to errors and be prejudicial to the interchangeability of documents.

Naturally, the quantitative parts of the data and the numbers identifying the geometrical elements which occur in the description of the work are expressed by figures.

When all the data have been written on the coding sheets, they are punched onto cards, or less commonly, onto punched tapes.

THE POST PROCESSOR

When the general program has been executed, the trajectory of the tool with respect to the part is known. The tape carrying this information is called the c.l. (cutter location) tape.

To generate the instructions necessary for the execution of a particular part on a given machine, account must not only be taken of the geometry of the part and the tools, but also of much more data relating to the machine and its control system.

The trajectory having been defined, the results of the computation are reprocessed for adaptation to the control requirements of a machine and a particular system; the program which performs the adaptation is often called the post processor.

A post processor is therefore exclusively intended for one combination of machine and system; it is much easier to program than a general system.

The principal points which a post processor must take into consideration are as follows.

Code

If the instructions intended for a machine are expressed in a code which is different from that of the computer, a transcoding is performed.

Coordinates

The results of the calculations are, in general, expressed exact to eight figures. This value must be rounded according to the precision of the machine. If, moreover, the length of the numbers or the position of the decimal point is prescribed, zeros must be added before, and/or after, the significant figures.

A conversion of units may also have to be made: inches to millimetres, degrees to grades or revolutions, minutes and seconds into ten-thousandths of a degree, etc.

Orientation of the axes

If the orientation of the axes is not standard, the sign of certain coordinates must be changed.

Shifting of the origin

When a machine has a fixed origin, the position of the origin of the part dimensions must be chosen in advance, and these must be corrected to take into account the corresponding translation.

At the same time a commutation of the axes may have to be made to turn the part through 90° if required by the characteristics of the machine.

Format

According to the type of format adopted, the directions may have to be added. If several points are situated on a line parallel to a plane of the coordinate system, one of their coordinates remains constant; in this case, some systems allow the omission of the coordinate from some blocks, and it is up to the post processor to take this into account.

Technological data

During the first phase, data, implicit or explicit (such as cutting speed, feed, coolant, etc.) which are relative to the conditions of work have not been taken into consideration.

They are utilized during the execution of the post processor. The range of speeds available is taken into account in rounding the theoretical values. If the machine is fitted with a manual control, these data are omitted from the punched tape, but they appear on the printout which acts as the instruction sheet for the operator.

Length of tools

The calculations made during the first phase only take into account the cutting edge of each tool. To work out the definitive instructions, account must be taken of the distance that separates the tip of each tool from the reference face of the spindle.

Capacity of the machine

The post processor checks that the dimensions of the part are compatible with the capacity of the machine assigned to its manufacture, i.e. it checks that no displacement exceeds the maximum travel.

Tool changing

According to the way in which the tool magazine and changer have been designed, the various operations may be accomplished in different ways. For example, if the magazine does not follow the displacements of the spindle head, the latter must be brought into a given position to make the exchange; if the changer has only one prong, the used tool must be replaced, and the next tool brought opposite the gripper, grasped and placed in the nose of the spindle. In the other case, if the changer has two grippers, one of them can grasp the next tool while the preceding one is still in use; as soon as the substitution is made, work resumes while the changer replaces the tool in the magazine and grasps the next.

Work cycles

The control system of the machine proper can itself contain sub-programs corresponding to various operations:

Drilling: rapid advance, slow feed, rapid return.
Tapping: rapid advance, tapping, withdrawal of the tap, rapid return.
Spot facing: rapid advance, slow feed, dwell, rapid return.
Deep drilling: rapid advance, slow feed, rapid return, rapid advance, slow feed, etc.

With some control systems, one block is sufficient to control the execution of a cycle.

If the system does not possess this facility, each cycle must be broken down, step by step.

Overshoot

When the axis of a milling cutter is to follow a rectangular contour ABC, precautions must be taken to avoid an overshoot at B which would substitute the trajectory AB′C for the planned one (Figure 6.5).

Taking into account the performance of the servo systems, a deceleration period is provided to prevent the tool axis from overshooting the position B by an amount greater than the permissible tolerance.

Also it is the post processor which establishes the sequences of acceleration and deceleration during rapid displacements.

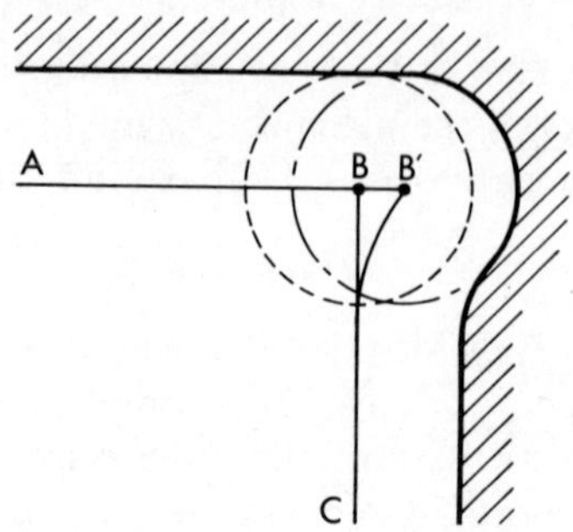

Fig. 6.5

Other tasks

The post processor can accomplish other tasks, such as the printing of instruction sheets, the calculation of the machining time for a part, the length of the tape required, the operator's pay slip, material order forms, etc.

CHECKING

During the preparation, writing, typing and execution of the program, several kinds of errors can creep in. Some are of human origin—the dimensioning of the drawing, the reading of dimensions, the writing of the program, punching of the cards; others are caused by a random or systematic malfunction of a typewriter, a punch or a computer.

A system must be protected automatically against certain errors. For example, if there has been substitution of a letter, the computer stops on meeting a word which is not in its vocabulary. The fault must then be corrected, and the computation resumed, causing additional expense and delays.

If an error has been made in a number, the program continues apparently normally, but the results are false. This can have serious consequences—injury to the operator, wear on the machine, rejection of a part.

There is always, at least, a loss of time and money, and multiple precautions must be taken to protect the system, as completely as possible, from the consequences of errors.

Some of these measures take place before the computer has started its work. This operation (computation) is very expensive and it must not be carried out aimlessly.

The execution of the post processor will hardly reveal an error of this order; what is checked, especially, in this phase of the work, is that the operations to be carried out are compatible with the capacities (dimensions, speed, power, etc.) of the machine to which they are being assigned.

Thus there is still the possibility, after the execution of the post processor, that errors remain, and a final check must then be made before machining is begun.

Program preparation

Data typing

On the programming sheet, the programmer only writes on every third line.

The operator punches the data by means of a typewriter which simultaneously prints, on the programming sheet, the text punched on cards or tape.

Each error by the operator is shown by a difference between the handwritten text and the one that is printed immediately below.

To detect a punching fault, the cards or tape are read on another typewriter, on which the programming sheet is mounted, and which prints, on a third line, what it has read from the cards or tape.

An error by the first machine is shown by a difference between the manuscript, or the first printing, and the second printing.

Double punching

The data is punched twice; a machine compares the results of these operations and indicates any difference.

There is little chance that the same error will be repeated exactly, especially if the precaution of having the work performed by two different operators is taken.

Successive double punching

Once the work has been finished, it is performed a second time by another operator; the machine reads, simultaneously, the results of the first operation, and stops itself as soon as it detects a difference between the two versions.

The operator is thus immediately alerted, and, after having checked which version is faulty, unlocks the machine and continues. Besides, if the error is caused by the operator, the machine automatically starts as soon as she has typed the correct character.

It can thus be made practically certain that the second punching is free from typing errors.

Plotting

If the program manuscript contains a numerical error, the preceding methods are almost inapplicable.

The punched tape or the magnetic tape intended for the machine tool is used to control the displacements of the stylus of a drafting machine; thus a diagram is obtained. Examination of the diagram can reveal an oversight or a gross error in a dimension, but is incapable of indicating a difference of a few hundredths of a millimetre.

Test piece

Despite the precautions taken, one cannot be certain of having avoided all errors, and it would thus be dangerous to commence the manufacture of a series of parts, or even of a single costly part.

If the parts are not very costly, the first part is submitted to a minute inspection, and the manufacture of the others is not begun until the first part has been accepted as satisfactory.

For costly parts, a dummy part is made first, perhaps from a rough part rejected because of a metallurgical fault, or even from soft wood.

6.5 EXECUTION OF THE JOB

Every effort must be made for the output of a numerically controlled machine to be as high as possible, i.e. for the elapsed time between the completion of a job and the commencement of the next to be reduced to a minimum.

While the machine is working automatically, the operator must have received everything necessary for the execution of the following job: the part, drawings, instruction sheets, punched tape and tools.

The tools are preset and tagged, either by an alphanumeric identification, or by being fitted with an index code.[12–15]

The operator can thus examine the part beforehand, study the drawings, and take note of the instruction sheet. If the machine has a tool magazine, he removes the tools which are no longer required, and replaces them by those that will be employed first in the next job.

If the parts must be clamped in advance on dummy tables, it is possible that the operator will not have time to carry out this task while supervising the machine in his charge. A specialized team performs this task, for all the machines, in a special area with the appropriate equipment: surface plates, blocks, gauges, dial gauges, etc.

The dummy tables have locating stops which are brought into contact with fixed stops on the machine table, so that the replacement of one part by the next requires neither adjustment nor measurement and, at the very most, only the modification of the offset of the origin.

Nor must inspection and handling of the parts slow down a process which has been speeded up by every other means.

The reliability of control systems almost totally eliminates checking of the result of the machining job, except for the first part in a series, but this checking is intended to correct a programming fault rather than to bring to light a control failure.

Handling is closely related to the organization of production; data processing, if it is properly organized, should avoid dead time between operations, which increases the delays in production and the number of parts in hand.

6.6 ARCHIVES

The records of a factory with numerically controlled machines can consist exclusively of a collection of punched tapes.

Two copies of the master tape relating to each part are retained. These tapes are not used to control a machine directly, but to produce automatically the instruction sheets, the pay slips, the order forms for material in stock and the control tapes for the machine, which only contain the instructions intended for the control system, to the exclusion of all the financial or administrative information listed on the master.

One of these tapes is never used, and is kept only to reproduce a new master tape if the other is damaged.

A clearly written label, stuck to the start of the tape, carries the name and number of the corresponding part, and allows its identification without having to read the coded perforations.

References

1. R. Le Grand, 'Preplanning for numerical control', *American Machinist*, 5 February 1962.
2. 'The economic and managerial aspects of numerically controlled machine tools', *Process Control and Automation*, April 1965.
3. E. J. Donelan, 'Planning for numerical control', *Machinery and Production Engineering*, 8 September 1965.
4. L. F. Hesse, 'How to get the most out of NC', *Production*, September 1965.
5. A. W. Young, 'N/C triggers management changes', *The Iron Age*, 27 January 1966.
6. L. B. Oldfield, 'Look before you leap into NC', *Metalworking Production*, 9 February 1966.
7. 'Getting started with NC', *American Machinist*, 7 November 1966.
8. 'Designed for tape controlled manufacture', *Production*, March 1964.
9. H. Soubiès-Camy, 'The foundations of programming for numerical control (Bases de la programmation en commande numérique)', *Mécanique et Électricité*, June 1968.
10. 'NC programming. Manual taping made easy. NC programming. Profile of a programmer', *Metalworking Production*, 28 June 1967.
11. H. Soubiès-Camy, 'Introduction to programming (Introduction à la programmation)'. C. Marchesseau, 'Point-to-point programming (La programmation point à point)'. M. Sachot, 'Practical examples of programming (Exemples pratiques de programmation)'. J. C. Morens, 'Automatic programming in continuous machining (La programmation automatique en usinage continu)'. R. Weill, 'New methods of programming (Méthodes nouvelles de programmation)'. *Bulletin du G.A.M.I.*, No. 3, 1967.
12. R. E. Green, 'Tape-controlled operations on business machine components', *Machinery*, 4 September 1963.
13. R. J. Klubertanz, 'Universal tooling for N/C', *American Machinist*, 17 February 1964.
14. P. A. Sidders, 'A machine tool builder makes extensive use of numerical control', *Machinery*, 7 April 1965.
15. G. R. Nelson, 'Preset NC tooling increases output at IBM', *Metalworking Production*, 22 June 1966.

Chapter 7

Conclusion

The advent of numerical control is much more than a technical event. It has brought about a complete transformation of the working methods used for the manufacture of small series parts.

With traditional methods, work preparation is relatively brief, and a large amount of it, such as the choice of cutting data and the sequence, is left to the initiative of the operator. Thus an important part of the operator's time is devoted to preparing his job and the duration of the effective working of his machine generally does not exceed 15 to 20 per cent of his work time.

In contrast, numerical control requires that the job should be minutely prepared, so that the machines operate almost without interruption.

To exploit all the possibilities in a logical and effective way, it is essential to depart from normal practices, and modify radically the distribution of personnel among the different professional categories, as well as the distribution of investments and working expenses under the various headings in the general budget.

There is no doubt that, more than all the apprehensions of a technical nature, it is this fundamental change in the way of thinking that has slowed down the development of numerical control, even in the extremely forward-thinking mechanical engineering industry of the United States.

In measuring this expansion, the statistics give only rather imprecise data. Some administration departments waited several years before distinguishing conventional machines from those equipped with numerical control; others have classified, under the same heading, machines whose size and value vary in the ratio of 50:1. Occasionally, no account has been taken of imports; machines whose displacements are controlled by preset dogs have been included as numerically controlled machines.

Lastly, it has not always been indicated whether the cost of the control systems has been included in the prices, and this can lead to a difference of between 15 and 50 per cent.

Taking into account all these reservations, some information characterizing the evolution of numerical control in the leading Western countries

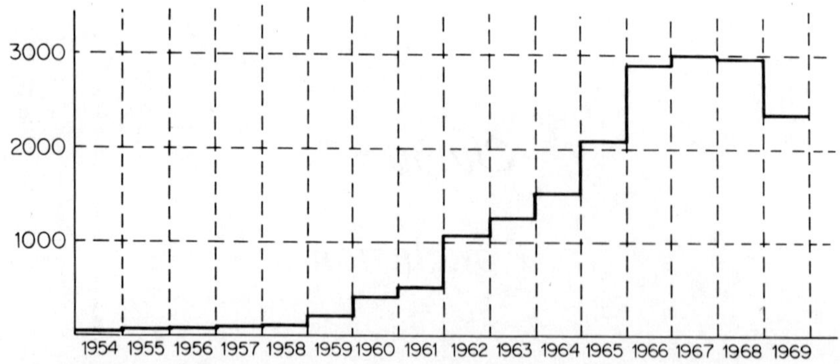

Fig. 7.1. Number of n.c. machines manufactured annually in the U.S.A.

will be found in the following graphs (Figures 7.1 to 7.7). Information concerning the Eastern and Far-Eastern countries is brief or non-existent.[1–9]

What forecasts can we hazard as to the development of numerical control?

On a technical level, the era of development is over, i.e. there are proven systems, which will cause the disappearance of the others, but this does not signify that systems that are more precise or economical than those in present use will not be invented.[10]

There are no doubts concerning the reliability of numerical control to make the eventual users hesitate; the area of its practical application is only bounded for economic or psychological reasons.

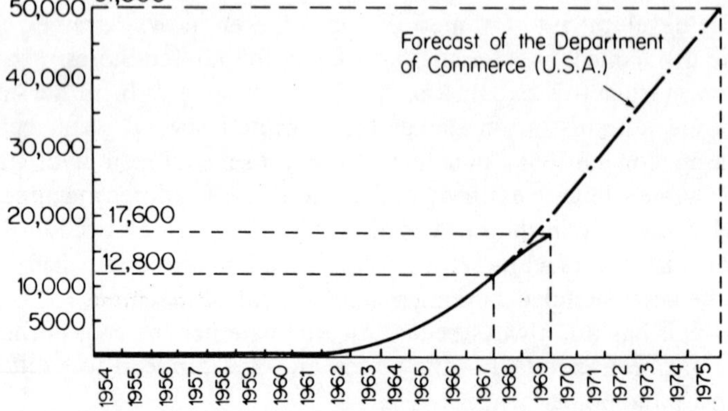

Fig. 7.2. Number of n.c. machines in use in the U.S.A.

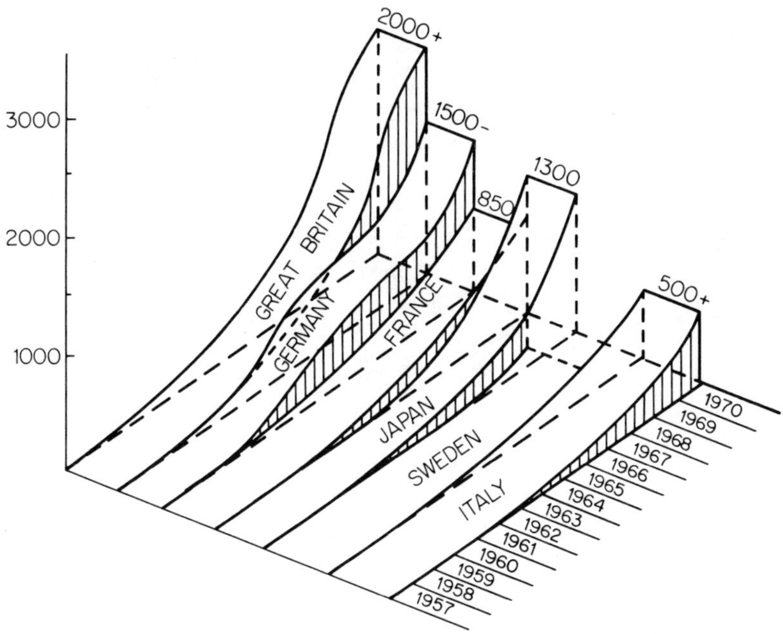

Fig. 7.3. Total number of n.c. machines in use in various countries

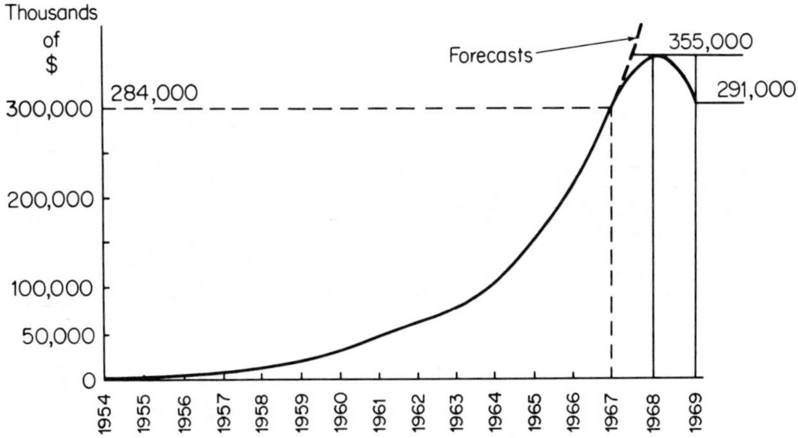

Fig. 7.4. Total value of n.c. machines produced annually in the U.S.A.

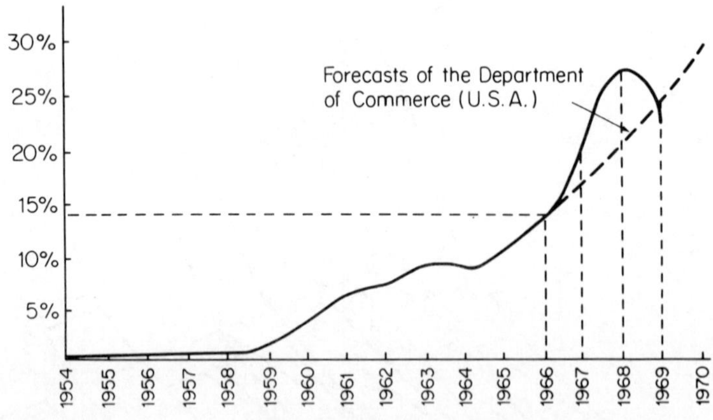

Fig. 7.5. Ratio of the value of n.c. machines to the total value of machine tools produced in the U.S.A.

For each type of part and for each type of operation, the area in which numerical control is economical ranges from the single part to runs which, taking into account the calculable and noncalculable advantages of this technique, number generally between 3000 and 10,000 parts. Beyond this, the use of conventional machines and apparatus becomes more profitable.

In the coming years, several reasons will certainly combine to extend this limit.

1. The cost of control systems will become lower.
2. The generalization of automatic programming will reduce the cost of the work preparation.

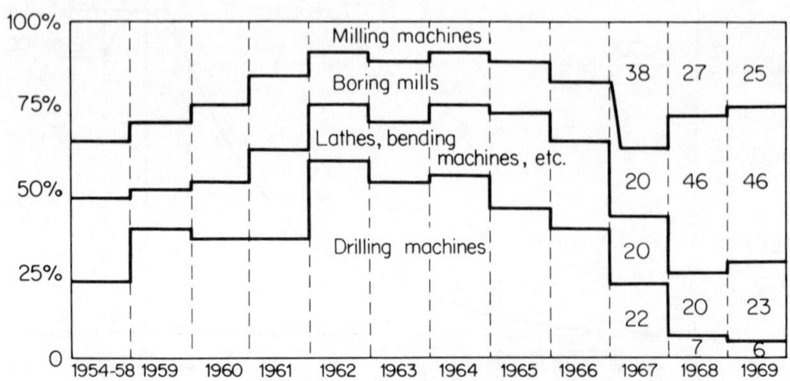

Fig. 7.6. N.C. machines in use in the U.S.A.; classification by type of machine. It is probable that machining centres are classified with boring machines.

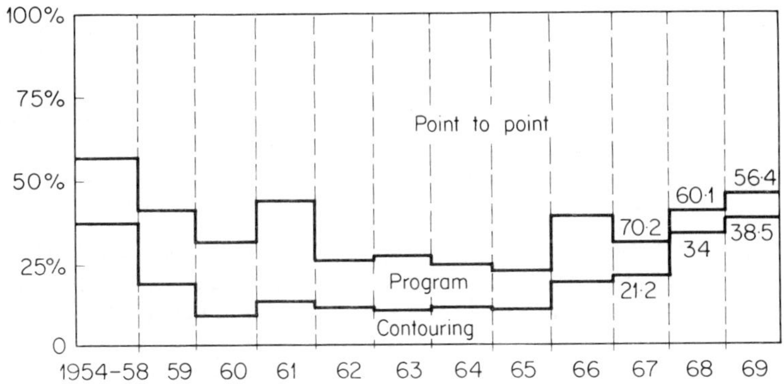

Fig. 7.7. N.C. machines in use in the U.S.A.; classification by type of control system

3. Teleprocessing, if its cost is sufficiently low, will reduce the lead times and the importance of work in hand.
4. Job organization will increase the output of machines and reduce, as a result, the hourly cost of their working.
5. The general rise in wages and the scarcity of specialists will necessitate their employment only in work preparation, and not in its execution.
6. When the requirements of numerical control are better known, the conception of parts will evolve so that the difficulties, and as a result the cost, of manufacture will tend towards a minimum.

On the other hand, we can legitimately doubt that numerical control will play a fundamental role in long series production, with the exception of operations such as balancing, sorting, matching, and matching weights (see p. 54), which must take into account the random characteristics of parts made at a high rate.

Since 1960, several transfer machines equipped with numerical control have been put into service.[11] They execute variations on the same type of part, and the program of each of their stations is modified taking into account the type of part which is fed to it; these changes have a bearing on the execution of the job, or the position of a unit with respect to the part; some multiple-head spindles are retractable.

In general, the different positions which one drilling unit has to take up are rather limited, and the number can scarcely exceed ten. In these circumstances, it may well be thought that several switches activated by independent dogs and linked to a conventional switch would perform, for a more modest cost, the same tasks as a numerical control system, which, on a 100 mm path, could define 20,000 different positions.

Is it possible to imagine the economical use of a series of numerically controlled monospindle machines instead of a transfer line? To answer this question, a comparison must be made, as an example, between two solutions relating to a part such as a cylinder block, which is quite well suited to machining by numerical control, with some simplifying assumptions.

Imagine a transfer line which is highly automated and capable of producing 150 parts per hour, covers 3000 m^2 and includes 600 tools.

If we add the building, the control equipment and the automatic-handling apparatus linking the machines and about 1500 tools carried on the spindles or in the sharpening shop or ready to be exchanged, the total value of the installation, is about £3 million ($7·2 million).

In the budget of this type of factory, the wages bill and the cost of power, tool steel, and various other factors are much less than the cost of amortizing the plant.

Moreover, this type of equipment is installed for a minimum period of seven to ten years. When its work is at an end, all the special components (fixtures, bush plates, multispindle heads, handling and storage apparatus etc.) have lost all their value, and it is assumed that the value of the rest (standard units, frames, etc.) just covers the costs of maintenance and reconditioning.

It can, therefore, very roughly be estimated, that the cost of machining a part is very nearly the proportion of the amortizement which is allocated to it.

A line working double shift for seven years at the rate of 150 parts per hour produces 4 million parts whose machining cost is roughly:

$$\frac{\text{£3 million}}{\text{4 million parts}} = \text{£0·75 (\$1·80) per part.}$$

On the other hand, if we assume that the cylinder block has dimensions of 400 × 320 × 220 mm, and that almost all its faces are to be milled or have oblique holes, five-axis machines costing more than £60,000 ($144,000) must be used to perform the same task, to which must be added the cost of a duplicate set of tools.

Taking into account the time for changing tools and setting, clamping and unclamping the parts, such machines cannot complete more than 200 operations an hour; they will thus complete a part in three hours.

Assuming the hourly rate to be equal to £3·75 ($9·00) per hour, the machining of a part will thus cost £11·25 ($27·00), to which must be added a proportion of the salary of an operator supervising three machines, with, say, charges of £0·90 ($2·20).

Thus, without taking into account the amortizement of the building, power, various supplies (lubricant, etc.) and the tool material, etc, we arrive at a total of £12·15 ($29·30).

This price is much higher than that of a part manufactured on a transfer line.

For the two systems to be similar, the latter would have to produce no more than 250,000 parts, which, reckoning at the medium rate, would only represent five months' production. It is obvious, therefore, that the hypothesis is absurd, since a manufacturer does not envisage equipping for this sort of work rate if he is not reasonably certain of using the plant over a period of seven to ten years.

It seems, therefore, that numerical control will never make serious inroads into the realm of special machines, and it may even be predicted that, between these two areas, another will continue to exist: the area of semi-specialized machines, equipped, for example, with multiple heads with movable spindles, or with readily changeable multiple heads, and which will be the most economic solution for runs lying between the two limits which we can approximately estimate at 20,000 and 500,000.

Some companies have already installed plant[12,13] in which the numerical system not only controls the machine tools, but also takes over the transport of parts and an important role in the preparation and management of the work.

Three principal activities may be distinguished:

preparation
execution (handling and machining)
management.

Work preparation is carried out, generally, by one shift working eight hours; it is subdivided into three parts:

(a) The preparation of the control instructions. This task is accomplished by the methods planners, with the help of a computer, and takes the physical form of punched or magnetic tapes.
(b) Setting and clamping of the parts. The parts are fixed on pallets whose dimensions, as well as reference points and fixing points, are standardized. This task is carried out in a specialized area of the factory where the work stations are equipped with apparatus for measuring, setting and securing the parts.
(c) Presetting of tools. After tools have been passed through the grinding shop, they are fitted to their holders, set by means of precision apparatus and identified. Then all those necessary for the machining of a given part are arranged in loaders and moved manually or automatically to the machines for which they are intended.

This work area is separate from the one where the machines linked by a handling system are grouped.

Detectors identify the parts carried on the pallets and direct them towards the machines that are equipped to machine them, i.e. those which are supplied with the relevant collections of tools and tapes.

As soon as a machine is free, it receives a new pallet and the work proceeds without interruption while the pallet carrying the newly machined part is sent by means of the general conveyor to the machine allocated for subsequent operations or, if the part is completed, to the inspection and unclamping station.

The management of the shop can be more or less automatic. The Sundstrand system (Rockford, Ill., U.S.A.) is organized for the manufacture of light-alloy gear cases having a certain similarity; each type is produced in a series of several dozen, or sometimes one or two hundreds. Several models are made simultaneously, and, each time a run is completed, a tool setter removes the control tapes and the corresponding tools, replacing them by those relevant to the next series.

Each machine is equipped to handle, with equal ease, four or five programs relating to the same part or to different parts. The distribution of work between the machines does not pose difficult problems. The superintendents undertake modifications of the program; the management is of a semiautomatic nature.

The Molins Machine Company has tackled a rather different problem, in manufacturing very varied parts made principally out of billets of light alloy, and including drilling, tapping, boring and two-dimensional contouring operations.[14]

The machining time of a part rarely exceeds a few minutes, and the daily production consists of a large variety of parts. Moreover, a great number of unforeseen eventualities can arise: shortage of material or tools, unavailability of a machine, or an urgent request for a spare part or a prototype.

A computer is informed of all these circumstances and knows, in addition, the manufacturing program, the availabilities and the execution times. It assigns the tasks to different machines, checks their work loads and modifies them in the light of events.

In this shop, the planning department works in a single shift, as do the operators who fix the parts on the pallets, but the machines and the management system work without interruption six days per week under the supervision of three lightly manned shifts.

This implementation marks an enormous step in the progress of automation of small series production, but to obtain this result the managers of Molins have had to accept a modification of shapes and material in the design of parts.

To complete its gearcase line, Sundstrand is setting up an even more automated plant, intended for the manufacture of bushes, pinions and shafts. It contains lathes, milling machines, drilling machines, gear-cutting machines, external and internal grinders and stations for induction treatment. All these machines are fitted with numerical-control systems and automatic apparatus which deposits the parts on the machine tools after taking them from baskets which carry them, and puts them back after the operation is finished.

The basket conveyor serves the machine shop, the control centre, the basket-loading and -unloading stations and their store.

Thus, there are today several shops where numerical control carries out almost all operations.

By means of a telecommunication system, it is possible to use a distant computer.

Numerical control, even extended to include the management of a shop, the preparation of programs, and the adaptive control of machines, would only occupy a small part of the time of a large computer.

Thus a computer normally intended for the execution of scientific computations or administrative work can also, provided it is not fully loaded, control machines or shops at some distance.[15–23]

This solution is only economical if the hourly cost of a large computer and of the communication system are sufficiently low. Moreover, the means of data transmission must be absolutely reliable.

It may be thought that for several more years systems controlled by a small computer will remain more economical than a system with a central computer; the latter will first come into its own where one is forced to use a powerful system for other tasks which do not fill it completely.

Programs which produce the process chart, the sequence of instructions and the execution time of a part from a dimensional drawing already exist. Obviously, they only apply as yet to relatively simple parts classed in a few well-defined groups, but, save for a genuine doubt about the economics of this undertaking, there is nothing to prevent this possibility from being extended to a much larger variety of parts.

In addition, there are known examples of systems which, as a function of their specifications, calculate the optimal dimensions of a mechanism or an assembly. Until now, the use of this method has been restricted to finding the principal dimensions of the framework of buildings[24,25] or lifting gear, the general characteristics of an electrical circuit, a turbine, an alternator or a transformer, or the dimensions of a multispindle head.[26]

All the objects thus defined have a considerable degree of similarity, but it would be an enormous undertaking to apply this method to parts when the methods of calculation involved would not by any means present the same characteristics of rigour, logic and relative simplicity.

To justify a similar effort, the problem must arise frequently, and it does not seem that a long-production-run industry, for example, could apply this solution in an economical way when it only creates new models at intervals of several years, and when the rhythm of secondary modifications is, at the very most, annual.

This thought is a slight reassurance to the technologist who, in the implementation of a project, thinks he is seeing the embryonic emergence of an inhuman world, in which a computer will have processed the results of an opinion poll which has been preconditioned by a skilful campaign of persuasion, will have synthetized the results of an inquiry on the future of a market, will then draft the specifications of a product, design the parts, determine their manufacturing process as well as the instructions for the machines, and finally drive the factory itself.

A novelist would add to this the rôle of measuring the efficiency and discipline of the few men who might possibly remain in this kind of system.

There is no doubt that, for a long time to come, the imagination and initiative of man will maintain his advantage over the power, speed, vigilance and rigour of machines.

All the practical applications which have been described in these chapters were born of the thought, the intuition and the efforts of men. Numerical control has multiplied the results, has taken charge of performing detailed tasks whose bounds have been fixed, whose rhythm has been imposed and whose direction has been charted by man. It will remain the servant, never raising itself to the level of the innovator.

All that we expect of it is that, having unloaded a great part of our physical work, it should leave to us the responsibility for imagining, choosing and deciding . . .

. . . And the time, our daily tasks completed, to enjoy a Shakespearian sonnet, a Handel oratorio or a Dürer engraving.

References

1. 'N/C progress in production plants', *Automation*, October 1963.
2. 'BDSA survey shows who buys numerical control', *Steel*, November 1963.
3. 'W. Germany counts her NC tools', *Metalworking Production*, 28 April 1965.
4. 'N/C: the market doubles', *The Iron Age*, 9 September 1965.
5. 'A study of the development of numerically controlled machines in the U.S.A. (Étude sur le développement des machines à commande numérique aux U.S.A.)', S.C.F.M.O., 150 Boulevard Bineau, 92-Neuilly, *U.S.A. Mission*, October 1964.
6. 'Where NC stands in the U.S.', *Metalworking Production*, 23 March 1966.
7. 'N/C revolution sweeps machine tool industry', *The Iron Age*, 7 July 1966.
8. T. M. Rohan, 'Machine tools: "sold out" for '67', *The Iron Age*, 5 January 1967.
9. F. R. J. Gruaz, 'The influence of numerical control on the structure and economy of a business. (Influence de la commande numérique sur la structure et l'économie de l'entreprise)', Cincinnati Milling Co., INSA meeting, 1968.
10. A. E. de Barr, 'Some thoughts on the future of numerical control', *Machinery and Production Engineering*, 20 April 1966.
11. B. C. Broscheer, 'NC for versatility in transfer machining', *Metalworking Production*, 15 March 1967.

12. 'Future NC will demand thorough process changes', *Steel*, 20 March 1967.
13. 'Variable mission machining', *American Machinist*, 9 September 1968.
14. D. T. N. Williamson, 'A new concept of manufacture (Une nouvelle conception de la fabrication)', *Ingénieurs et Techniciens*, September 1967.
15. D. A. Bell and P. M. Will, 'Some principles and circuit techniques for controlling machine tools from a central digital computer', *Journal Brit. I.R.E.*, August 1963.
16. 'Computer to control NC machines', *Automation*, February 1967.
17. 'Computers make the big difference', *Metalworking Production*, 15 March 1967.
18. 'Numerical control: the new ideas', *American Machinist*, 27 March 1967.
19. 'The computer in N/C's future', *The Iron Age*, 8 June 1967.
20. 'Direct computer control of machine tools', *Machinery and Production Engineering*, 20 December 1967.
21. 'Direct NC from a computer', *American Machinist*, 25 March 1968.
22. R. W. Murray, 'NC line links computer to tools', *The Iron Age*, 22 February 1968.
23. 'Omnicontrol system for on-line computer control of n.c. machines', *Machinery and Production Engineering*, 10 April 1968.
24. J. Foucriat, Possibilities of computer use in metallic fabrication (Possibilités de calcul électronique en construction métallique)', *Arts et Métiers*, April 1965.
25. 'Computer frees engineers to think big', *Steel*, 10 December 1962.
26. 'How Cross use special computer in designing special machines', *American Machinist*, 12 November 1966.
27. 'The computer in NC's future', *The Iron Age*, 8 June 1967.
28. 'NC penetration into metalcutting shipments', *The Iron Age*, 10 September 1970.
29. G. Garratt, 'NC machine tools under direct computer control (details of the Sundstrand Omnicontrol system, and some applications in the U.S.A.)', *Machinery and Production Engineering*, 2 July 1969.
30. 'New computer numerical control system', *Machinery*, April 1970.
31. C. H. Wick, 'Economical direct computer control for machine tools', *Machinery and Production Engineering*, 24 June 1970.

Appendix 1

Spline Curves

(by A. R. Forrest, University of Cambridge Computer Laboratory)

Physical splines

For many years the salient curves of ships, cars and aircraft have been drafted manually in lofts by means of splines. Splines are thin metal or wood laths which are bent elastically round weights or 'ducks' placed at points chosen to give the desired curve (Figure Al.1).

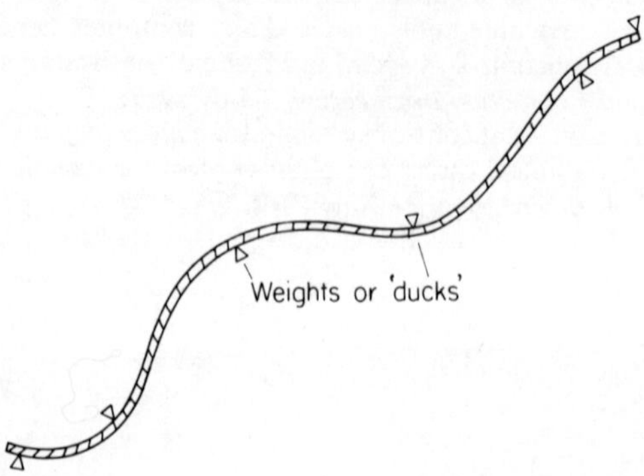

Fig. A1.1

Mathematical splines

Because the use of physical splines is well established and the resulting curves are acceptable, there is good reason to employ mathematically analogous curves in describing shapes to be cut by numerically controlled machines. The physical spline assumes a shape for which its internal strain energy is a minimum. Mathematically, this can be expressed by requiring that the integral

$$B \int K^2 \, ds \qquad \text{(A1.1)}$$

is a minimum when integrated over the entire length of the spline. B is a constant expressing the stiffness of the spline, K is the curvature of the spline and s is the arc length ($0 \leq s \leq L$, where L is the total length of the spline).

In Cartesian-coordinate terms we can rewrite the integral:

$$\int \frac{y''^2 \, \mathrm{d}x}{(1+y'^2)^{\frac{5}{2}}}. \tag{A1.2}$$

As this integration is difficult to perform, it is usual to assume that the slope of the spline, $y' = \mathrm{d}y/\mathrm{d}x$, is everywhere sufficiently small to be ignored (i.e. $y' \ll 1$). The problem then reduces to minimizing the integral:

$$\int y''^2 \, \mathrm{d}x. \tag{A1.3}$$

A notable exception to this approach is the KURGLA curve-smoothing algorithm developed by Mehlum for the AUTOKON ship-designing system.[1] Mehlum performs his approximation at a later stage in the algorithm, and thus the problem of avoiding curves with large slopes does not arise.

Solving the variational problem of minimizing expression A1.3 subject to the spline passing through specified points or 'knots', it is found that the resultant curve—the mathematical spline—is a series of cubic curves joined piecewise at the knots. The curves are continuous in position, slope and curvature at the knots. In practice, the mathematical spline is a good approximation to the physical spline, provided that the slopes are not too large. If the slopes are to be large, as occurs frequently in practical cases, there are two choices. A local set of axes can be chosen for each arc of the spline, usually along, and perpendicular to, the chord of the arc. Alternatively, we may use parametric (i.e. vector) methods so that we compute a spline which is a function of t for each component of the vector $\mathbf{P}(t)$. This method has the advantage of being axis independent. It also permits the computation of fully three-dimensional or twisted curves, but, strictly speaking, these curves do not behave as three-dimensional elastica because torsional terms in the strain-energy minimization have been ignored.

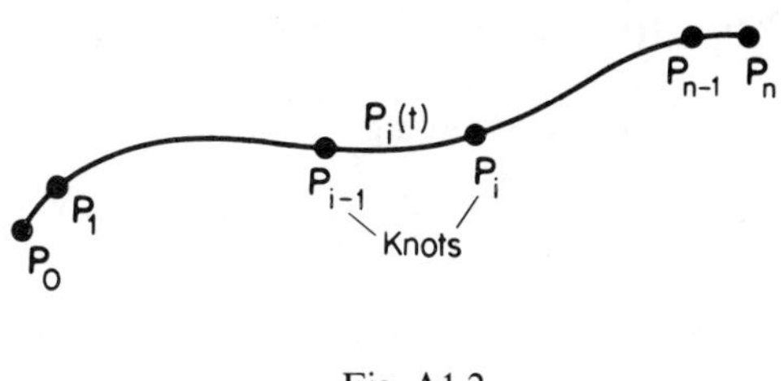

Fig. A1.2

Spline computation

It has already been shown in Chapter 5 that a sequence of cubic arcs continuous in position, slope and curvature can be constructed through a given set of points. However, the step-by-step or marching technique described in Chapter 5 tends, in practice, to be unstable numerically, and error accumulates as the computation proceeds. The error shows as unwanted oscillations in the curve. A different formulation of these splines enables the computation to be executed in a more stable manner.

Consider the parametric spline whose ith span $\mathbf{P}_i(t)(0 \leq t \leq 1)$ runs from $\mathbf{P}_i(0) = \mathbf{P}_{i-1}$ to $\mathbf{P}_i(1) = \mathbf{P}_i$ (Figure A1.2). Then

$$\mathbf{P}_i(1) = \mathbf{P}_{i+1}(0) = \mathbf{P}_i, \text{ etc.} \tag{A1.4}$$

We may write the equation of the span in the form:

$$\mathbf{P}_i(t) = \mathbf{P}_{i-1}(1-3t^2+2t^3)+\mathbf{P}_i(3t^2-2t^3)+\mathbf{P}'_{i-1}(t-2t^2+t^3) + \mathbf{P}'_i(-t^2+t^3). \tag{A1.5}$$

The values of the tangent vectors $\mathbf{P}'_i$ are unknown; we can compute them by making use of the continuity conditions.

Differentiating equation A1.5:

$$\mathbf{P}'_i(t) = \mathbf{P}_{i-1}(-6t+6t^2)+\mathbf{P}_i(6t-6t^2)+\mathbf{P}'_{i-1}(1-4t+3t^2) + \mathbf{P}'_i(-2t+3t^2) \tag{A1.6}$$

and again:

$$\mathbf{P}''_i(t) = \mathbf{P}_{i-1}(-6+12t)+\mathbf{P}_i(6-12t)+\mathbf{P}'_{i-1}(-4+6t)+\mathbf{P}'_i(-2+6t). \tag{A1.7}$$

Hence:

$$\mathbf{P}''_i(0) = -6\mathbf{P}_{i-1}+6\mathbf{P}_i-4\mathbf{P}'_{i-1}-2\mathbf{P}'_i \tag{A1.8}$$

and

$$\mathbf{P}''_i(1) = 6\mathbf{P}_{i-1}-6\mathbf{P}_i+2\mathbf{P}'_{i-1}+4\mathbf{P}'_i. \tag{A1.9}$$

If the span $\mathbf{P}_i(t)$ is to be continuous in second derivative with the span $\mathbf{P}_{i+1}(t)$ at the point $\mathbf{P}_i$, then, from equations A1.8 and A1.9, we must have:

$$6\mathbf{P}_{i-1}-6\mathbf{P}_i+2\mathbf{P}'_{i-1}+4\mathbf{P}'_i = -6\mathbf{P}_i+6\mathbf{P}_{i+1}-4\mathbf{P}'_i-2\mathbf{P}'_{i+1}.$$

Hence, dividing by 2 and rearranging:

$$\mathbf{P}'_{i-1}+4\mathbf{P}'_i+\mathbf{P}'_{i+1} = -3\mathbf{P}_{i-1}+3\mathbf{P}_{i+1}. \tag{A1.10}$$

If the parameter of each span runs from $i-1$ to i instead of 0 to 1, the same recurrence relation may be obtained. If, however, the parameter interval

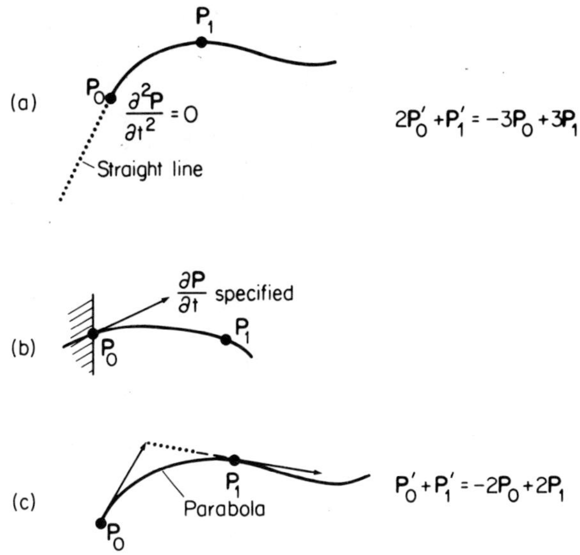

Fig. A1.3. End conditions

varies from span to span, say $L_{i-1} \leq t \leq L_i$ for the ith span, where L_i is the cumulative chord length of the polygon joining the knots of the spline and $l_i = L_i - L_{i-1}$ is the chord length of the ith span, the recurrence relation becomes:

$$\frac{1}{l_i}\mathbf{P}'_{i-1} + \left(\frac{2}{l_i} + \frac{2}{l_{i+1}}\right)\mathbf{P}'_i + \frac{1}{l_{i+1}}\mathbf{P}'_{i+1} = \frac{3}{l_i^2}\mathbf{P}_{i+1} + \left(\frac{3}{l_i^2} - \frac{3}{l_{i+1}^2}\right)\mathbf{P}_i + \frac{3}{l_{i+1}^2}\mathbf{P}_{i+1}. \quad \text{(A1.11)}$$

The relationships A1.10 and A1.11 can only be applied to the $n-1$ internal points $\mathbf{P}_1$ to $\mathbf{P}_{n-1}$ of the spline, and, to complete the solution, we need two additional conditions.

The most common additional conditions are (Figure A1.3):

(a) Free end(s) [no bending moment at the end of the spline, i.e. zero curvature at the end(s)].

$$\mathbf{P}''_0 = 0, \qquad \mathbf{P}''_n = 0.$$

From equation A1.8, setting $i = 1$, we have:

$$\mathbf{P}''_1(0) = \mathbf{P}''_0 = 0 = -6\mathbf{P}_0 + 6\mathbf{P}_1 - 4\mathbf{P}'_0 - 2\mathbf{P}'_1.$$

Rearranging:

$$2\mathbf{P}'_0 + \mathbf{P}'_1 = -3\mathbf{P}_0 + 3\mathbf{P}_1 \quad \text{(A1.12)}$$

and, similarly, for the other end of the spline:

$$\mathbf{P}'_{n-1}+2\mathbf{P}'_n = -3\mathbf{P}_{n-1}+3\mathbf{P}_n. \tag{A1.13}$$

(b) Specified first derivative(s) at the end(s) (i.e. the spline is 'built-in', or cantilevered). For convenience, we write this

$$\mathbf{P}'_0 = \mathbf{T}_0 \tag{A1.14}$$

$$\mathbf{P}'_n = \mathbf{T}_n \tag{A1.15}$$

(c) Parabolic end span(s). From equation A1.5 for $i = 1$, if $\mathbf{P}_1(t)$ is to be a parabola, no cubic terms should exist. Therefore

$$\mathbf{P}'_0+\mathbf{P}'_1 = -2\mathbf{P}_0+2\mathbf{P}_1. \tag{A1.16}$$

Similarly, for the other end of the spline

$$\mathbf{P}'_{n-1}+\mathbf{P}'_n = -2\mathbf{P}_{n-1}+2\mathbf{P}_n. \tag{A1.17}$$

Using the $n-1$ equations A1.10 and two additional equations we can construct the matrix equation

$$A\mathbf{P}'_i = B\mathbf{P}_i \tag{A1.18}$$

where A is an $(n+1)\times(n+1)$ matrix and B is a matrix with $(n+1)$ rows. In the case where a tangent vector is specified at a knot, the column vector $\mathbf{P}_i$ (strictly speaking a vector of vectors) also includes the specified tangent vector. For example, a four-span spline with one end built in and the other free has the matrix equation:

$$\begin{bmatrix} 1 & 0 & 0 & 0 & 0 \\ 1 & 4 & 1 & 0 & 0 \\ 0 & 1 & 4 & 1 & 0 \\ 0 & 0 & 1 & 4 & 1 \\ 0 & 0 & 0 & 1 & 2 \end{bmatrix} \begin{bmatrix} \mathbf{P}'_0 \\ \mathbf{P}'_1 \\ \mathbf{P}'_2 \\ \mathbf{P}'_3 \\ \mathbf{P}'_4 \end{bmatrix} = \begin{bmatrix} 1 & 0 & 0 & 0 & 0 & 0 \\ 0 & -3 & 0 & 3 & 0 & 0 \\ 0 & 0 & -3 & 0 & 3 & 0 \\ 0 & 0 & 0 & -3 & 0 & 3 \\ 0 & 0 & 0 & 0 & -3 & 3 \end{bmatrix} \begin{bmatrix} \mathbf{T}_0 \\ \mathbf{P}_0 \\ \mathbf{P}_1 \\ \mathbf{P}_2 \\ \mathbf{P}_3 \\ \mathbf{P}_4 \end{bmatrix} \tag{A1.19}$$

and a similar spline with one end free and a parabolic constraint at the other has the matrix equation:

$$\begin{bmatrix} 2 & 1 & 0 & 0 & 0 \\ 1 & 4 & 1 & 0 & 0 \\ 0 & 1 & 4 & 1 & 0 \\ 0 & 0 & 1 & 4 & 1 \\ 0 & 0 & 0 & 1 & 1 \end{bmatrix} \begin{bmatrix} \mathbf{P}'_0 \\ \mathbf{P}'_1 \\ \mathbf{P}'_2 \\ \mathbf{P}'_3 \\ \mathbf{P}'_4 \end{bmatrix} = \begin{bmatrix} -3 & 3 & 0 & 0 & 0 \\ -3 & 0 & 3 & 0 & 0 \\ 0 & -3 & 0 & 3 & 0 \\ 0 & 0 & -3 & 0 & 3 \\ 0 & 0 & 0 & -2 & 2 \end{bmatrix} \begin{bmatrix} \mathbf{P}_0 \\ \mathbf{P}_1 \\ \mathbf{P}_2 \\ \mathbf{P}_3 \\ \mathbf{P}_4 \end{bmatrix}. \tag{A1.20}$$

We solve for the tangent vectors $\mathbf{P}'_i$ by inverting the matrix A and pre-multiplying B by the inverse:

$$\mathbf{P}'_i = A^{-1}B\mathbf{P}_i. \tag{A1.21}$$

Each span of the spline is then specified using equation A1.5. The matrix A is tridiagonal and consequently may be inverted rapidly and accurately.

Expressions similar to equations A1.12 and A1.13 can be obtained for splines parameterized according to chord length.

Use of splines

The mathematical theory of splines is explained in considerable detail by Ahlberg, Nilson and Walsh and others.[2–4] Spline curves are normally used to represent the kind of curves found in ship hulls and car bodies. They provide a smooth yet flexible curve, particularly for so-called sculptured shapes [spline techniques were used by Ferguson to obtain the tangent vectors for his curves and surfaces (Reference 16, Chapter 5)]. A spline curve has the property that the entire curve changes if one knot is moved. Thus the effects of changes cannot be localized within a spline. Splines should not be used where there is a definite discontinuity in a second derivative, as, for instance, with a straight line joining a circular arc.

References

1. E. Mehlum, 'Curve and surface fitting based on a variational criteria for smoothness', Central Institute for Industrial Research, Oslo, Norway, December 1969.
2. J. H. Ahlberg, E. N. Nilson and J. L. Walsh, *The theory of splines and their applications*, Academic Press, New York, 1967.
3. E. N. Nilson, 'Cubic splines on uniform meshes', *Comm. ACM*, **13**, No. 4, April 1970.
4. T. N. E. Greville, '*Theory and applications of spline functions*', Academic Press, New York, 1969.

Appendix 2

Coons Surfaces

(by A. R. Forrest, University of Cambridge Computer Laboratory)

It is profitable to consider Coons surfaces (Reference 20, Chapter 5) in more detail than in Chapter 5 and to relate them to other methods of surface definition such as UNISURF and Ferguson's surfaces. To do so we must adopt a somewhat different notation from that used in Chapter 5 for Coons surfaces.

Patch defined by boundary curves

Basically, a Coons surface is defined by its boundary conditions. A patch has four sides (although these may be degenerate, as with UNISURF). Let a point on a surface patch be denoted by $\mathbf{P}(u, v)$ where $0 \leq u, v \leq 1$. The boundary curves of the patch are $\mathbf{Q}(0, v)$, $\mathbf{Q}(1, v)$, $\mathbf{Q}(u, 0)$ and $\mathbf{Q}(u, 1)$ (Figure A2.1), and these curves intersect at the four corner points $\mathbf{Q}(0, 0)$, $\mathbf{Q}(0, 1)$, $\mathbf{Q}(1, 0)$ and $\mathbf{Q}(1, 1)$. We use $\mathbf{Q}$ to denote a vector which is used to *define* a surface and, in distinction, $\mathbf{P}$ to denote a vector which is a property of the surface *defined by* the vector conditions $\mathbf{Q}$.

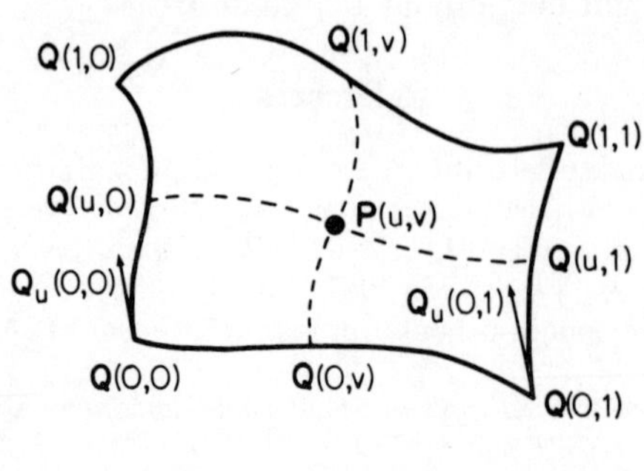

Fig. A2.1. Coons surface defined by boundary curves only

If we let $i = 0$ or 1 and $j = 0$ or 1, the four boundary curves and the four corner points may be written: $\mathbf{Q}(i, v)$, $\mathbf{Q}(u, j)$ and $\mathbf{Q}(i, j)$.

The simplest Coons surface patch is the patch defined by its boundary curves alone:

$$\begin{aligned}\mathbf{P}(u, v) = {} & \mathbf{Q}(0, v) \,.\, \mathrm{F}_0(u) + \mathbf{Q}(1, v) \,.\, \mathrm{F}_1(u) + \mathbf{Q}(u, 0) \,.\, \mathrm{F}_0(v) + \mathbf{Q}(u, 1) \,.\, \mathrm{F}_1(v) \\ & - \mathbf{Q}(0, 0) \,.\, \mathrm{F}_0(u) \,.\, \mathrm{F}_0(v) - \mathbf{Q}(0, 1) \,.\, \mathrm{F}_0(u) \,.\, \mathrm{F}_1(v) \\ & - \mathbf{Q}(1, 0) \,.\, \mathrm{F}_1(u) \,.\, \mathrm{F}_0(v) - \mathbf{Q}(1, 1) \,.\, \mathrm{F}_1(u) \,.\, \mathrm{F}_1(v)\end{aligned} \tag{A2.1}$$

or, using the i, j notation as a shorthand,

$$\mathbf{P}(u, v) = \mathbf{Q}(i, v) \,.\, \mathrm{F}_i(u) + \mathbf{Q}(u, j) \,.\, \mathrm{F}_j(v) - \mathbf{Q}(i, j) \,.\, \mathrm{F}_i(u) \,.\, \mathrm{F}_j(v).$$

Each corner point is subtracted from the equation of the surface because these points already occur twice in the equation; they each lie on two of the boundary curves.

The functions $\mathrm{F}_i(u)$ and $\mathrm{F}_j(v)$ are called 'blending functions'; they serve to ensure that the surface patch has the desired boundary conditions [i.e. $\mathbf{P}(i, v) \equiv \mathbf{Q}(i, v)$, etc.] and also to control continuity with adjoining surface patches. If we require only positional continuity with another patch, i.e. that the patch should, in fact, be bounded by $\mathbf{Q}(i, v)$ and $\mathbf{Q}(u, j)$, the following conditions apply to the blending functions:

$$\mathrm{F}_i(k) = \delta_{ik}, \qquad \mathrm{F}_j(k) = \delta_{jk} \tag{A2.2}$$

where δ_{ik} and δ_{jk} are Kronecker deltas, $i, j, k = 0$ or 1, i.e.

$$\begin{aligned}\mathrm{F}_i(k) &= 1 \text{ if } i = k \\ &= 0 \text{ if } i \neq k.\end{aligned}$$

There are few restrictions on the form of the blending functions: they should be continuous, and preferably monotonic, for $0 \leq u, v, \leq 1$. The simplest forms are:

$$\mathrm{F}_0(t) = 1 - t, \qquad \mathrm{F}_1(t) = t. \tag{A2.3}$$

We can verify that the surface patch goes through the boundary curves by setting $u, v = 0, 1$ in equation A2.1; if adjoining patches share common boundary curves, they will be continuous in position with the patch.

There are, equally, few constraints on the boundary curves; they should intersect at the patch corners and should, in most circumstances, be continuous.

If we use the more complex blending functions, as suggested in Chapter 5:

$$\begin{aligned} F_0(t) &= 1-3t^2+2t^3 \\ F_1(t) &= 3t^3-2t^3 \end{aligned} \tag{A2.4}$$

we note that $F_0(0) = 1$, $F_0(1) = 0$, etc., and also that

$$F'_0(0) = F'_0(1) = F'_1(0) = F'_1(1) = 0.$$

Now

$$\begin{aligned} \frac{\partial \mathbf{P}(u, v)}{\partial u} &= \mathbf{P}_u(u, v) \\ &= \mathbf{Q}(i, v)F'_i(u) + \mathbf{Q}_u(u, j)F_j(v) - \mathbf{Q}(i, j)F'_i(u)F_j(v) \end{aligned}$$

hence

$$\begin{aligned} \mathbf{P}_u(0, v) &= \mathbf{Q}(i, v)F'_i(0) + \mathbf{Q}_u(0, j)F_j(v) \\ &\quad - \mathbf{Q}(i, j)F'_i(0)F_j(v) \\ &= \mathbf{Q}_u(0, j)F_j(v) = \mathbf{Q}_u(0, 0)F_0(v) + \mathbf{Q}_u(0, 1)F_i(v) \end{aligned} \tag{A2.5}$$

and similarly

$$\mathbf{P}_u(1, v) = \mathbf{Q}_u(1, j)F_j(v)$$

i.e.

$$\mathbf{P}_u(i, v) = \mathbf{Q}_u(i, j)F_j(v). \tag{A2.6}$$

Thus the tangent vector across the boundary is determined in this case by the tangent vectors across the boundary at the ends of the boundary (Figure A2.1). Continuity with other patches under these conditions is described in Chapter 5, and higher-order continuity may be obtained by using more powerful blending functions (i.e. higher order in the polynomial case). However, we are asking rather much by using boundary curves and blending functions alone to specify slope and higher-order continuity between patches; experience has shown that this leads to patches which have unwanted flat regions. What we can do is to use a more flexible form of Coons surface. As was mentioned at the beginning, a surface patch may be specified by its *boundary conditions.*

Patch with defined boundary curves and slopes

Suppose we wish to specify not only the boundary curves $\mathbf{Q}(i, v)$ and $\mathbf{Q}(u, j)$, but also the cross-boundary slopes $\mathbf{Q}_u(i, v)$ and $\mathbf{Q}_v(u, j)$. These cross-boundary slopes must have the properties:

$$\left.\frac{d\mathbf{Q}_u(i, v)}{dv}\right|_{v=j} \equiv \left.\frac{d\mathbf{Q}_v(u, j)}{du}\right|_{u=i} \equiv \left.\frac{\partial^2\mathbf{Q}(u, v)}{\partial u\partial v}\right|_{\substack{u=i\\v=j}} \equiv \mathbf{Q}_{uv}(i, j) \equiv \mathbf{P}_{uv}(i, j). \quad (A2.7)$$

These vectors are called 'twist vectors' by Coons.

We now wish to construct a surface satisfying the boundary conditions $\mathbf{Q}(i, v)$, $\mathbf{Q}(u, j)$, $\mathbf{Q}_u(i, v)$ and $\mathbf{Q}_v(u, j)$ (Figure A2.2).

Consider the equation

$$\begin{aligned}\mathbf{P}(u, v) &= [F_i(u)G_i(u)]\begin{bmatrix}\mathbf{Q}(i, v)\\ \mathbf{Q}_u(i, v)\end{bmatrix} + [\mathbf{Q}(u, j)\mathbf{Q}_v(u, j)]\begin{bmatrix}F_j(v)\\ G_j(v)\end{bmatrix} \quad (A2.8)\\ &\quad - [F_i(u)G_i(u)]\begin{bmatrix}\mathbf{Q}(i, j) & \mathbf{Q}_v(i, j)\\ \mathbf{Q}_u(i, j) & \mathbf{Q}_{uv}(i, j)\end{bmatrix}\begin{bmatrix}F_j(v)\\ G_j(v)\end{bmatrix}\\ &= \mathbf{Q}(i, v)F_i(u) + \mathbf{Q}_u(i, v)G_i(u)\\ &\quad + \mathbf{Q}(u, j)F_j(v) + \mathbf{Q}_v(u, j)G_j(v) - \mathbf{Q}(i, j)F_i(u)F_j(v)\\ &\quad - \mathbf{Q}_u(i, j)G_i(u)F_j(v) - \mathbf{Q}_v(i, j)F_i(u)G_j(v)\\ &\quad - \mathbf{Q}_{uv}(i, j)G_i(u)G_j(v).\end{aligned}$$

$F_i(u)$ and $F_j(v)$ are the blending functions we have encountered earlier, but we impose additional conditions on them. As well as $F_i(k) = \delta_{ik}$, $F_j(k) = \delta_{jk}$, we must have $F_i'(k) = 0$, $F_j'(k) = 0$ for $k = 0$ or 1 in order that the boundary curves should not interfere with the cross-boundary slopes we have specified.

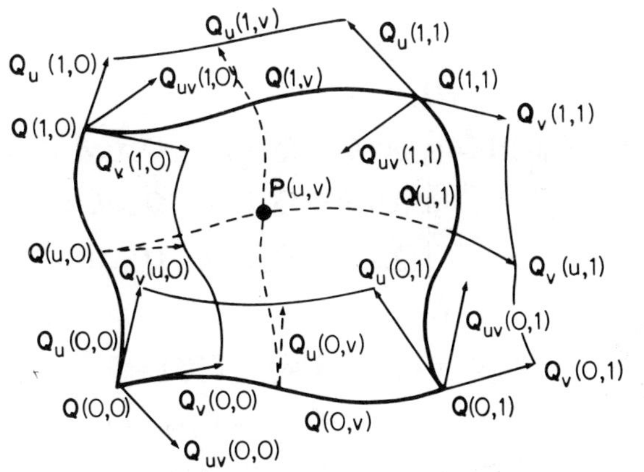

Fig. A2.2. Coons surface defined by boundary curves and boundary slopes only

$G_i(u)$ and $G_j(v)$ are also blending functions. In order that the cross-boundary slopes should not affect the boundary curves, we must impose the conditions $G_i(k) = G_j(k) = 0$ for $k = 0, 1$. In order that the surface patch should have the required cross-boundary slopes,

$$G_i'(k) = \delta_{ik}$$

and

$$G_i'(k) = \delta_{jk}.$$

The simplest blending functions obeying these conditions are (as before)

$$\begin{aligned} F_0(t) &= 1-3t^2+2t^3 \\ F_1(t) &= 3t^2-2t^3 \\ G_0(t) &= t-2t^2+t^3 \\ G_1(t) &= -t^2+t^3. \end{aligned} \tag{A2.9}$$

The easiest way to ensure slope continuity between adjacent patches is to ensure that they have both a common boundary curve and a common cross-boundary slope.

By specifying the cross-boundary slope of the patch as well as the boundary curves, more control is obtained over the shape of the surface.

Higher-order boundary conditions

Likewise we may specify patches with defined cross-boundary second derivatives:

$$\begin{aligned} \mathbf{P}(u, v) = {} & [F_i(u)G_i(u)H_i(u)] \begin{bmatrix} \mathbf{Q}(i, v) \\ \mathbf{Q}_u(i, v) \\ \mathbf{Q}_{uu}(i, v) \end{bmatrix} + [\mathbf{Q}(u, j)\mathbf{Q}_v(u, j)\mathbf{Q}_{vv}(u, j)] \begin{bmatrix} F_j(v) \\ G_j(v) \\ H_j(v) \end{bmatrix} \\ & -[F_i(u)G_i(u)H_i(u)] \begin{bmatrix} \mathbf{Q}(i, j) & \mathbf{Q}_v(i, j) & \mathbf{Q}_{vv}(i, j) \\ \mathbf{Q}_u(i, j) & \mathbf{Q}_{uv}(i, j) & \mathbf{Q}_{uvv}(i, j) \\ \mathbf{Q}_{uu}(i, j) & \mathbf{Q}_{uuv}(i, j) & \mathbf{Q}_{uuvv}(i, j) \end{bmatrix} \begin{bmatrix} F_j(v) \\ G_j(v) \\ H_j(v) \end{bmatrix} \end{aligned} \tag{A2.10}$$

where

$$\begin{aligned} F_i(k) &= \delta_{ik} \qquad & F_i'(k) &= F_i''(k) = 0 \\ G_i'(k) &= \delta_{ik} \qquad & G_i(k) &= G_i''(k) = 0 \\ H_i''(k) &= \delta_{ik} \qquad & H_i(k) &= H_i'(k) = 0 \end{aligned} \tag{A2.11}$$

and so on for any order of boundary conditions.

The bicubic patch

There is one particular form of the Coons patch which is widely used—the so called bicubic patch. Suppose that the blending functions are the cubic polynomials given earlier, and that the boundary curves and cross-boundary slopes are also cubic functions. Suppose, moreover, that the boundary curves are specified in the form:

$$\begin{aligned} \mathbf{Q}(i, v) &= \mathbf{Q}(i, j)\mathrm{F}_j(v) + \mathbf{Q}_v(i, j)\mathrm{G}_j(v) \\ \mathbf{Q}(u, j) &= \mathbf{Q}(i, j)\mathrm{F}_i(u) + \mathbf{Q}_u(i, j)\mathrm{G}_i(u) \end{aligned} \tag{A2.12}$$

and the cross-boundary slopes are defined similarly:

$$\begin{aligned} \mathbf{Q}_u(i, v) &= \mathbf{Q}_u(i, j)\mathrm{F}_j(v) + \mathbf{Q}_{uv}(i, j)\mathrm{G}_j(v) \\ \mathbf{Q}_v(u, j) &= \mathbf{Q}_v(i, j)\mathrm{F}_i(u) + \mathbf{Q}_{uv}(i, j)\mathrm{G}_i(u). \end{aligned} \tag{A2.13}$$

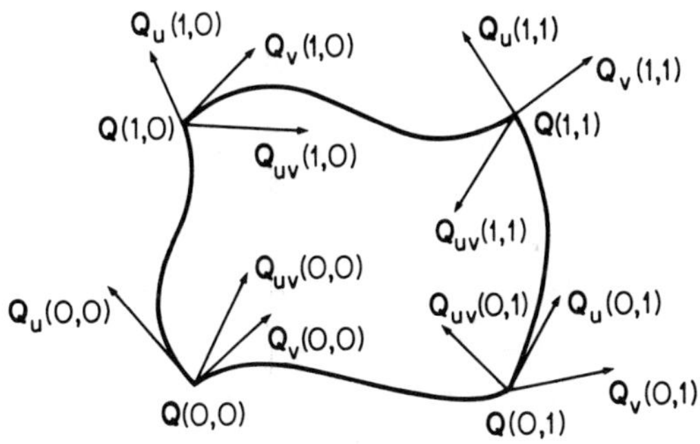

Fig. A2.3. The bicubic Coons path

We can now simplify the equation of the patch (Figure A2.3)

$$\mathbf{P}(u, v) = [\mathrm{F}_i(u)\mathrm{G}_i(u)]\begin{bmatrix} \mathbf{Q}(i, j) & \mathbf{Q}_v(i, j) \\ \mathbf{Q}_u(i, j) & \mathbf{Q}_{uv}(i, j) \end{bmatrix}\begin{bmatrix} \mathrm{F}_j(v) \\ \mathrm{G}_j(v) \end{bmatrix} \tag{A2.14}$$

or, in the form it is usually written by Coons,

$$\mathbf{P}(u, v) = [u^3 \quad u^2 \quad u \quad 1]\begin{bmatrix} 2 & -2 & 1 & 1 \\ -3 & 3 & -2 & -1 \\ 0 & 0 & 1 & 0 \\ 1 & 0 & 0 & 0 \end{bmatrix}$$

$$\times \begin{bmatrix} \mathbf{Q}(0,0) & \mathbf{Q}(0,1) & \mathbf{Q}_v(0,0) & \mathbf{Q}_v(0,1) \\ \mathbf{Q}(1,0) & \mathbf{Q}(1,1) & \mathbf{Q}_v(1,0) & \mathbf{Q}_v(1,1) \\ \mathbf{Q}_u(0,0) & \mathbf{Q}_u(0,1) & \mathbf{Q}_{uv}(0,0) & \mathbf{Q}_{uv}(0,1) \\ \mathbf{Q}_u(1,0) & \mathbf{Q}_u(1,1) & \mathbf{Q}_{uv}(1,0) & \mathbf{Q}_{uv}(1,1) \end{bmatrix}$$

$$\times \begin{bmatrix} 2 & -3 & 0 & 1 \\ -2 & 3 & 0 & 0 \\ 1 & -2 & 1 & 0 \\ 1 & -1 & 0 & 0 \end{bmatrix} \begin{bmatrix} v^3 \\ v^2 \\ v \\ 1 \end{bmatrix}. \tag{A2.15}$$

Equivalence with other patch definitions

This patch is specified by sixteen vectors; we can thus compare it with any other bicubic surface definition. For example, for a bicubic UNISURF

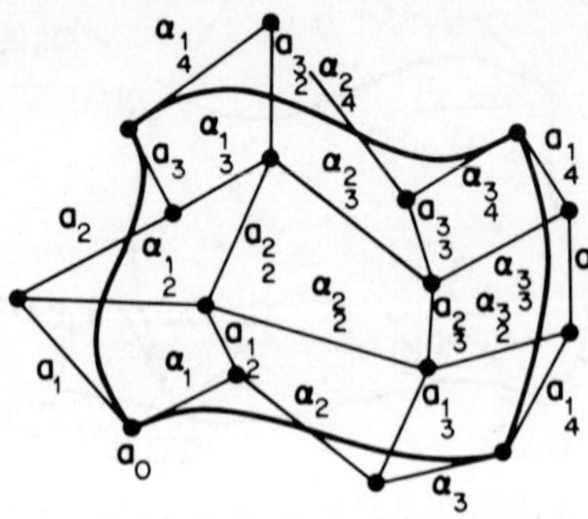

Fig. A2.4. The equivalent UNISURF patch mesh

surface patch (Figure A2.4), we can rewrite equations 5.75 and 5.76:

$$\mathbf{P}(u,v) = [u^3 \quad u^2 \quad u \quad 1] \begin{bmatrix} 0 & 1 & -2 & 1 \\ 0 & -3 & 3 & 0 \\ 0 & 3 & 0 & 0 \\ 1 & 0 & 0 & 0 \end{bmatrix}$$

$$\times \begin{bmatrix} \mathbf{a}_0 & \boldsymbol{\alpha}_1 & \boldsymbol{\alpha}_2 & \boldsymbol{\alpha}_3 \\ \mathbf{a}_1 & (\mathbf{a}_{\substack{1\\2}} - \mathbf{a}_{\substack{1\\1}}) & (\mathbf{a}_{\substack{1\\3}} - \mathbf{a}_{\substack{1\\2}}) & (\mathbf{a}_{\substack{1\\4}} - \mathbf{a}_{\substack{1\\3}}) \\ \mathbf{a}_2 & (\mathbf{a}_{\substack{2\\2}} - \mathbf{a}_{\substack{2\\1}}) & (\mathbf{a}_{\substack{2\\3}} - \mathbf{a}_{\substack{2\\2}}) & (\mathbf{a}_{\substack{2\\4}} - \mathbf{a}_{\substack{2\\3}}) \\ \mathbf{a}_3 & (\mathbf{a}_{\substack{3\\2}} - \mathbf{a}_{\substack{3\\1}}) & (\mathbf{a}_{\substack{3\\3}} - \mathbf{a}_{\substack{3\\2}}) & (\mathbf{a}_{\substack{3\\4}} - \mathbf{a}_{\substack{3\\3}}) \end{bmatrix}$$

$$\times \begin{bmatrix} 0 & 0 & 0 & 1 \\ 1 & -3 & 3 & 0 \\ -2 & 3 & 0 & 0 \\ 1 & 0 & 0 & 0 \end{bmatrix} \begin{bmatrix} v^3 \\ v^2 \\ v \\ 1 \end{bmatrix} \tag{A2.16}$$

$$= [u^3 \quad u^2 \quad u \quad 1] \begin{bmatrix} 0 & 1 & -2 & 1 \\ 0 & -3 & 3 & 0 \\ 0 & 3 & 0 & 0 \\ 1 & 0 & 0 & 0 \end{bmatrix}$$

$$\times \begin{bmatrix} \mathbf{a}_0 & \boldsymbol{\alpha}_1 & \boldsymbol{\alpha}_2 & \boldsymbol{\alpha}_3 \\ \mathbf{a}_1 & (\underset{2}{\boldsymbol{\alpha}_1} - \underset{1}{\boldsymbol{\alpha}_1}) & (\underset{2}{\boldsymbol{\alpha}_2} - \underset{1}{\boldsymbol{\alpha}_2}) & (\underset{2}{\boldsymbol{\alpha}_3} - \underset{1}{\boldsymbol{\alpha}_3}) \\ \mathbf{a}_2 & (\underset{3}{\boldsymbol{\alpha}_1} - \underset{2}{\boldsymbol{\alpha}_1}) & (\underset{3}{\boldsymbol{\alpha}_2} - \underset{2}{\boldsymbol{\alpha}_2}) & (\underset{3}{\boldsymbol{\alpha}_3} - \underset{2}{\boldsymbol{\alpha}_3}) \\ \mathbf{a}_3 & (\underset{4}{\boldsymbol{\alpha}_1} - \underset{3}{\boldsymbol{\alpha}_1}) & (\underset{4}{\boldsymbol{\alpha}_2} - \underset{3}{\boldsymbol{\alpha}_2}) & (\underset{4}{\boldsymbol{\alpha}_3} - \underset{3}{\boldsymbol{\alpha}_3}) \end{bmatrix}$$

$$\times \begin{bmatrix} 0 & 0 & 0 & 1 \\ 1 & -3 & 3 & 0 \\ -2 & 3 & 0 & 0 \\ 1 & 0 & 0 & 0 \end{bmatrix} \begin{bmatrix} v^3 \\ v^2 \\ v \\ 1 \end{bmatrix}. \tag{A2.17}$$

We can thus transform readily from the Coons form to the UNISURF form by equating corresponding coefficients in the bicubic expansion.

Twist vectors

If we deliberately set all the twist vectors $\mathbf{Q}_{uv}(i, j)$ to be zero, the bicubic patch is then identical with Ferguson's biparametric surface. If, in addition, $u = x$ and $v = y$, so that $\mathbf{P}(u, v)$ is the vector $[x\ y\ z(x, y)]$, then $z(x, y)$ is a surface similar to the Inaba surface described in Chapter 5.

Unless we are prepared to use a large number of patches, or do not mind a lack of fairness, it is dangerous to set all the twist vectors to zero, because this causes local flattening at patch corners. However, it is rather difficult to specify twist vectors directly, and it is better to derive them by equating the Coons surface with a comparable definition such as UNISURF, where the vectors defining the surface have an effect on surface shape which is more readily comprehended by a non-mathematician.

INDEX